12

In Defense of Self

How the Immune System Really Works

In Defense of Self
How the Immune System Really Works

William R. Clark

OXFORD
UNIVERSITY PRESS
2008

OXFORD

UNIVERSITY PRESS

Oxford University Press, Inc., publishes works that further
Oxford University's objective of excellence
in research, scholarship, and education.

Oxford New York
Auckland Cape Town Dar es Salaam Hong Kong Karachi
Kuala Lumpur Madrid Melbourne Mexico City Nairobi
New Delhi Shanghai Taipei Toronto

With offices in
Argentina Austria Brazil Chile Czech Republic France Greece
Guatemala Hungary Italy Japan Poland Portugal Singapore
South Korea Switzerland Thailand Turkey Ukraine Vietnam

Copyright © 2008 by Oxford University Press, Inc.

Published by Oxford University Press, Inc.
198 Madison Avenue, New York, New York 10016
www.oup.com

Oxford is a registered trademark of Oxford University Press

Library of Congress Cataloging-in-Publication Data
Clark, William R., 1938–
In defense of self : how the immune system really works / William R. Clark.
  p. ; cm.
Includes index.
ISBN: 978-0-19-533663-4
ISBN: 978-0-19-533555-2 (pbk.)
1. Immunology. 2. Immunity. 3. Immunologic diseases. I. Title.
[DNLM: 1. Immune System-physiology. 2. Immune System-physiopathology.
3. Immunotherapy. QW 504 C596i 2007]
QR181.7.C5532 2007
616.07'9—dc22   2006021047

Printed in the United States of America
on acid-free paper

Preface

The immune system—it is the only thing standing between us and a sea of microbial predators that could send us to an early and ugly death. Our world is filled with invisible microorganisms that find the human body a delightful place to live and rear a family. Our insides are not only rent free, but warm, moist, full of nutrients, and protected from the elements. Who could ask for more?

It is the job of the immune system to make certain this invasion doesn't happen. Oh, we do let in a few microbes that we put to work helping us digest food or process vitamins, but the vast majority of potentially disease-causing (pathogenic) microbes—bacteria, viruses, molds, and a few parasites—are kept at bay. And this same system would also be our only defense during the early hours of a bioterrorist attack, which might employ these very same microbes.

But the immune system isn't perfect. For one thing, it prevents us from accepting potentially life-saving organ transplants. It is also capable of overreacting; turning too much force against foreign invaders, whatever the source; and causing serious—occasionally lethal—collateral damage to our tissues and organs. Worse yet, our immune systems may decide that we *ourselves* are foreign and begin snipping away at otherwise healthy tissues, resulting in autoimmune disease. And finally, the immune system is itself the target of one of the most deadly viruses humans have ever known: HIV, the agent of AIDS.

How does this incredible system work? Honed over millions of years of evolutionary selection to keep us alive in a biological mine-field, the immune system has developed an impressive armamen-tarium of powerful chemical and cellular weapons that make short work of hostile viruses and bacteria. It also has evolved amazing genetic strategies to keep pace with invading microbes that can reproduce—and thus alter their own genetic blueprint—in under an hour.

Knowing just how the immune system functions has been key to some of the most important medical advances of the past hun-dred years, from the development of vaccines to the treatment of allergies, autoimmunity, and cancer, from prolonging organ trans-plants to combatting AIDS. Once the exclusive province of highly skilled specialists, this information can now be laid out in an en-gaging and informative story, accessible to all.

This book will take you on a tour of your immune system and show you how it works and the brilliant strategies that have been developed to keep us alive until we have had a chance to repro-duce—and with a little luck, a few years beyond that. You will not only gain a better understanding of how an important part of your own body works, but you will also be able to tune into the exciting research themes of today that will produce the medi-cal breakthroughs of tomorrow. And if you want to delve further into any of the topics discussed in this book, just e-mail me at wclark222@cs.com.

Enjoy!

William Clark, PhD
Los Angeles, 2007

Acknowledgments

I would like to thank Annemarie Shibata, Edwin L. Cooper, Thomas Valente, and Shinji Kasahara for reading the manuscript at various stages of production. Their insightful comments and suggestions have made this a much better book than I could have produced on my own.

I would also like to thank Celine Park for her help with the illustrations in this book.

William Clark, PhD
Los Angeles, 2007

Contents

PART I

HOW THE IMMUNE SYSTEM WORKS

What Is an Immune System?

You may not know it, but your immune system is huge. It's second in size only to your liver. But since it's spread throughout your body (Figure 1.1), you are probably only minimally aware of it. Lymph nodes are everywhere. The immune system has its own organs, like the thymus and the spleen. Its cells permeate your entire bloodstream. Your tonsils and adenoids and appendix are all part of the immune system, and so is the inside of your bones! In just a moment we'll look at many of these components of the immune system, how they work, and how they fit together.

But first, here's a simple question you may have asked yourself.

WHY DO I NEED AN IMMUNE SYSTEM?

You have an immune system for one reason and one reason only. In its absence, the human body would be a delightful place for microorganisms like bacteria, viruses, funguses, and parasites to live and raise their families. Your body is warm, wet, and chock full of the nutrients microbes need to survive and reproduce their own kind. Compare that with other places microbes are known to live: boiling sulfur vents at the bottom of the sea, for example, or beneath the frozen Arctic tundra. But don't expect gratitude from the microbes that have taken up residence in your body. The vast majority of them frankly don't care what happens to you. Some of them can make you very sick. Some can kill you. And that could interfere with nature's plan for you—to survive and reproduce *your* own kind.

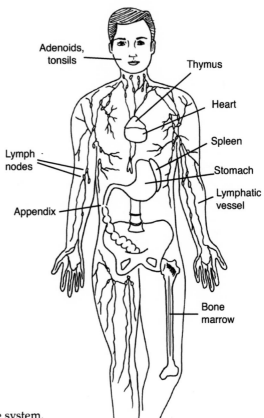

FIGURE 1.1
The human immune system.

Microbes were around long before humans. In the beginning, all life forms on earth consisted of just a single cell, and today many of them, such as the bacteria and the fungi, are still just that—single cells. (Viruses are a special case—they are not really cells at all, and are not really "alive." We'll discuss these vicious little bioids in detail in chapter 5.) When certain of these single-cell microbes evolved into more complex multicelled plants and animals, it wasn't long before some of the more clever ones who stayed behind gave up living in the hostile environment provided by the still-young earth and took up residence inside their multicellular descendents. They became parasites.

So multicellular life forms had to evolve in an environment filled with tiny would-be invaders and assassins. In order to do this, complex organisms had to develop systems to protect themselves from their invaders long enough to reproduce. Fortunately for us, they did! And by the way, humans are not at all unique in having immune systems. Every multicellular organism on this planet—both plant and animal—faced the same problem and had to evolve some sort of microbial defense system.

The microbes did not just sit back and let larger animals and plants build defense systems they could not penetrate and take advantage of. The story of evolution of larger plants and animals is to a great extent the story of coevolution of larger organisms with their microbial adversaries. For every defense strategy developed by multicellular organisms to ward off microbes, the microbes developed counterstrategies to evade them. In turn, their hosts were forced to develop new defense mechanisms or perish.

AN UNEVEN RACE. . .

Microbes have one distinct advantage in this race to survive: speed of reproduction. One of the dominant themes in evolution is size. Over evolutionary time, animals generally became larger. There are lots of reasons for this, but one of the most obvious is that the bigger you are, the more likely you are to be a predator rather than someone else's prey. But of course, the bigger you are, the longer it takes to put you together from scratch. A bacterium, if it can stay warm and get enough to eat and drink, can reproduce itself in less than an hour. We take a dozen years or so at the very least.

The reason this is important is that evolutionary changes, in response to environmental threats like extinction, result from changes in an evolving organism's DNA—**mutations** in its genes that give it a reproductive advantage over its brothers and sisters and cousins and aunts. And the major source of these mutations comes into play when DNA is reproduced during the generation of offspring.

Every time a cell divides, the DNA inside the "parent" cell has to be copied to produce DNA for "daughter" cells. This copying is fairly precise; it has to be if the offspring is going to be a viable copy of the parent. But it is not *absolutely* precise. Mistakes are made during copying of DNA. Most of these **copy errors** are edited out. But the editing process is also not perfect, and in every generation a small number of changes creep into DNA—into genes. These slight variations of genes between generations provide the raw stuff of evolution and natural selection.

What does this have to do with the competition for dominance and survival, versus submission and death, between single-cell microbes and more complex organisms such as ourselves? Don't microbes make mistakes, too?

Yes, but consider this. If a fertilized human egg and a bacterium were entered into a race to see who could make more cells, in the first three days of the nine months it takes to make a newborn human, the bacterium—which, remember, can double every half-hour or so—would have produced enough copies of itself to equal in mass all of the human beings now alive on the face of the earth! Of course, this would never happen, because the bacterium and its descendants could never get enough food to keep the reproductive process going.

But you see the problem. Microbes can generate the genetic changes that drive their evolution trillions of times faster than human beings. So larger organisms like us had to develop a repertoire of tricks to keep up with the microbes' incredible reproductive (and mutational) pace. Immune systems of the type we possess, which came into evolutionary existence with the vertebrates (fish, amphibians, reptiles, birds, and mammals like us), have developed a means of recognizing and destroying not only every microbe that exists in the world today, but also any microbe that might ever evolve any time in the future, whether we had seen anything like it before or not. We'll see exactly how we clever mammals do this in the next chapter, but now let's take a look at how the immune system is put together inside our bodies.

WHAT DOES IT TAKE TO MAKE A MAMMALIAN IMMUNE SYSTEM?

As we said, even the earliest multicellular animals and plants on earth had some sort of defense system to ward off invasion by microbial predators. As living things became more complex, so did their defense mechanisms. Some of these mechanisms were successful enough to have been carried forward all the way into human beings. They form a part of our defenses against microbes called **innate immunity**. So your immune system contains some components that have been around a billion years or more, in addition to the more recent mechanisms found only in vertebrates.

For many years, scientists assumed that innate immunity was just a quaint reminder of an earlier time and a cruder immune system no longer critical to our survival. Nothing could be further from the truth. We could not survive without it. The problem is, it is no longer sufficient to protect us completely. We will discuss innate immunity, and how it works, when we discuss details of the immune response to microbes in chapter 5.

The Bone Marrow

A good place to begin a description of the mammalian immune system is the **bone marrow** (Figure 1.2). This is the pale yellow-ish-white, jelly-like substance found in the center of most bones in your body. The function of bone marrow is to give rise to all of the cells found in the blood. Most of these are **red blood cells** ("erythrocytes"), which carry oxygen and give blood its color. But scat-

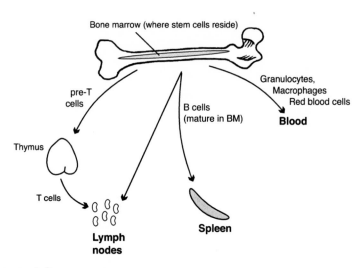

FIGURE 1.2
Bone marrow stem cells give rise to all cells found in blood.

tered among these red blood cells floating around our bodies are so-called **white blood cells** ("leukocytes"), which are the foot soldiers of the immune system. The sole job of white blood cells is to protect against invasion by microbial predators. All red blood cells are alike, but what we refer to collectively as white blood cells is actually an assortment of many different types of white cells, each with a different and important immune function.

Both red and white blood cells derive ultimately from something called a **bone marrow stem cell**. The official name of this stem cell is the **hematopoietic stem cell**, meaning a stem cell that gives rise to cells of the blood. The stem cells of the bone marrow do not themselves have any of the characteristics of mature blood cells, but something intriguing happens to stem cells when they are triggered to divide. Normally, when a cell divides, it produces two identical daughter copies of itself. But when stem cells divide, they do so asymmetrically—they produce one daughter that is an exact copy of themselves and a second daughter that now is poised to take off from the stem cell parent and become something else. In the case of hematopoietic stem cells, one daughter is an exact copy of the stem cell, and the other daughter is a cell that proceeds down a pathway leading to one of the blood cell types—a red cell, or one of the many types of white cells we'll meet shortly.

The hematopoietic stem cell is the critical cell in a bone marrow transplant. It can give rise to every cell of the blood, but these transplants can be very dangerous, as we will see in chapter 9.

The Thymus

The thymus is a glandular organ that lies just above the heart. In gourmet kitchens around the world the thymus from calves and lambs can, with appropriate skill, be turned into something called sweetbreads. But in our bodies it is the site for development of a special immune cell called a **T cell**, the "T" reflecting its thymic origin. These cells actually arise from stem cells in bone marrow, but they leave the bone marrow via the blood before they are fully mature and ready to help protect us against microbial invaders (Figure 1.2). The period of thymic maturation is particularly important, because it is where T cells learn what is self in the body and what is not. Failure to make this distinction can result in the immune system turning against self, which results in **autoimmunity** (chapter 12). The thymus reaches its maximal size during adolescence, and declines gradually thereafter.

Lymphocytes

One of the types of white blood cells arising from bone marrow stem cells is called **lymphocytes**. There are two major subsets of lymphocytes: **T cells** and **B cells**. T cells, as we have just seen, get their name from the fact that they must pass through the thymus to complete their maturation. The intense selection they undergo within the thymus results in the death of at least 90% of the T cells arising from stem cells in the bone marrow. These are presumably the T cells that could potentially react with self components. B cells arise in, and complete their entire maturation in, the bone marrow. The job of B cells is to produce a blood protein called an **antibody**, which hunts down and helps destroy foreign invaders swimming around in body fluids. T cells do not make antibody. Rather, they promote an itchy, painful process called **inflammation**, which provides a powerful defense against all sorts of microbial invaders. We will look at this process in detail in chapter 5. T cells also help B cells make antibody. As we will see, one highly specialized T cell, called a **killer T cell**, can seek out and physically destroy cells in the body determined to be harboring viruses or other intracellular parasites.

There are a number of other key players in the white cell repertoire that play important roles in our response to foreign invaders, but rather than describe them all here, we will look at them more closely when we encounter them in the real-life situations they were designed to deal with. The two lymphocytes we just met—T and B cells—together provide one of the most important distinctions between the immune systems of vertebrates like us and the innate defense systems of animals that came earlier. They provide us with a spectacularly precise, highly nuanced, and incredibly effective defense system that earthworms could only dream of.

Lymph Nodes and Spleen

When T cells and B cells reach full maturity, they leave the thymus and bone marrow, respectively, and migrate to and take up residence in the **lymph nodes** and **spleen** (Figure 1.2). You have only one spleen, a large, red organ just next to the stomach, but you have literally hundreds of small lymph nodes scattered throughout the body. These structures are not just tiny sacs full of cells, however. Lymph nodes have a strict internal structure that is repeated in every node (Figure 1. 3). B cells are found mostly in

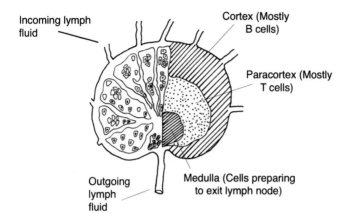

Incoming lymph fluid

Cortex (Mostly B cells)

Paracortex (Mostly T cells)

Outgoing lymph fluid

Medulla (Cells preparing to exit lymph node)

FIGURE 1.3
A human lymph node.

the outer (cortex) region of each node; T cells reside in a region called the paracortex.

As blood and lymph pass through these structures, the substances they are carrying are trapped there and examined. Potentially interesting materials are also brought to lymph nodes from around the body by a special scavenging cell called a **dendritic cell**. All of this is examined closely by T and B cells. Anything that is "self" is allowed to pass through. Anything that is "not self" triggers a series of alarms and sets an immune response in motion, activating T cells and B cells that recognize the offending antigen as foreign. B cells, once activated, begin to make and secrete the protein called antibody, which we will look at in detail in the next two chapters. The activated T cells leave the lymph nodes, going on patrol in the body to seek out the source of the problem. When they find it, they organize an attack that results in clearance of the offending material from the body.

The spleen has several functions. One of them is to remove dead and dying red cells from the blood and recover the iron from the hemoglobin they carry. But portions of the spleen also function like a giant lymph node and trap foreign material for inspection by resident T and B cells.

What Is This Stuff Called Lymph?

The blood does more than just carry red and white blood cells around the body. It also carries digested food and oxygen to all of the body's tissues. These foodstuffs are dissolved in the liquid part of the blood (**plasma**) and are unloaded through the very tiniest branchings of blood vessels called **capillaries**. The red cells unload their oxygen and pick up carbon dioxide; the foodstuffs are absorbed by nearby cells, which discharge waste products from previous feedings into the surrounding cell-bathing fluids.

A one-way flow of liquid from blood to tissue would quickly dry up your circulation, and the tissues of your body would be a soggy, bloated mess. The body's method of delivery of food and oxygen creates a gigantic plumbing problem, which it solved by developing a **lymphatic drainage system** (Figure 1.1). Where there are capillaries in the body (and that is *everywhere*), you will find a series of drains and drainpipes called **lymphatic sinuses and lymphatic vessels**. These are similar to veins, although somewhat more fragile. They drain the excess fluid from around cells and tissues and shunt it back to the blood circulation. Whereas blood vessels break up into smaller and smaller branches and eventually become capillaries, the lymphatic vessels start out tiny and merge with one another into a series of major trunks that eventually empty **lymph fluid** back into the bloodstream at the great veins of the neck.

Completely aside from its function in maintaining the plumbing integrity of the body, mammals have cleverly co-opted the lymphatic draining system for use by the immune system. It is along the scattered lymphatic network that the lymph nodes are stationed. Remember, many things that travel through the blood end up in the lymphatic circulation, so the various lymphatic branchings and trunks are ideal places to position lookout posts to keep an eye on traffic. The lymph nodes are themselves also served by blood vessels, so it is difficult for anything traveling around the body in either blood or lymph to escape surveillance in the lymph nodes. The spleen, although not receiving lymph fluids, still acts as an effective filter for foreign materials in the blood.

Whenever you get a cut or other wound, microbes and other potentially harmful material can cross into tissue spaces, where they are quickly swept into the general lymphatic traffic and trickle through downstream lymph nodes. In the case of cancer, cells escaping from a tumor into surrounding tissue fluids are also likely to be trapped in a nearby lymph node. That is why in the case of many cancers, surgeons collect nearby surrounding lymph nodes for examination by a pathologist. The presence or absence of cancer cells in these nodes is an important factor in planning treatment strategies.

Now that you've had a brief introduction to the basic structure of the immune system, let's take a look at how it works. There's still more to learn about the system itself, but we can pick that up as we go along.

2

Antibodies

MUCKING ABOUT IN THE DARK...

Humans knew they had an immune system centuries before they had any idea what an immune system was. In the fifth century B.C. Thucydides, in his *History of the Peloponnesian War*, declared that soldiers, who had been made sick by and then recovered from one of the many diseases rolling through army camps, were ideal nurses for freshly stricken soldiers. This was so, he said, since it was known that recovered soldiers *"never caught the same disease twice, and if they did, it was never fatal."* Without knowing it, Thucydides was describing the phenomenon of immunological **memory**. It's what we mean when we say we're "immune" to something. By the end of this chapter, you will know exactly what that means.

At the end of the 1700s, A British surgeon-apothecary named Edward Jenner found that by purposely exposing people—even children—to cowpox, a mild disease of cows related to the deadlier smallpox in humans, he could protect them against subsequent exposure to full-blown smallpox, which would normally be fatal in about one-third of those getting the disease. These are not the type of experiments physicians are encouraged to dabble in today, but at the time—possibly because he chose to practice on members of the working classes—no one seemed to mind. It is from these experiments using material from a cow, the Latin name for which is *vacca*, that we derive the term **vaccination**. (Vaccinations today have nothing to do with cows.)

Neither Jenner nor anyone else had the slightest idea why this worked, but being practical men rather than philosophers, they didn't really care. In Jenner's day up to half a million people died each year from smallpox in Europe alone, and three to four times that many were disfigured for life by deeply scarred pit marks on their face and elsewhere. The practice of vaccination was fully embraced by the British army, and it gave their soldiers a decided health advantage some decades later when they fought against their irregularly vaccinated American cousins.

One of the problems in figuring out how vaccination worked was that no one really understood what caused disease in the first place. Most assumed it was just rotten luck. Clergymen liked the idea that disease was one way God punished people for their sins. It was nearly a century after Jenner before scientists discovered the existence of microbes and proved their involvement in a wide range of what came to be known as **infectious diseases**. It took a while to convince people that living things they could not see could wreak such havoc.

But eventually the scientists proved their point. The higher incidence and rapid spread of many diseases in crowded cities, as opposed to rural farms, suggested that catching a disease probably involved people actually touching one another. That, in turn, implied some sort of physical entity passing between people, maybe from clothes or skin-to-skin contact. The practice of isolating (**quarantining**) infected people from the general population, which seemed to work in many cases, also supported this idea.

A LIGHT AT THE END OF THE TUNNEL

The first proof that an invisible entity—a microbe, in fact—could indeed cause disease came not with infectious diseases in humans, but from attempts to deal with the silkworm blight in midnineteenth-century France. Silkworm growers would see a few worms get sick as they chewed away on mulberry leaves. Within a few weeks, the entire herd was toast! Many thought a poison had somehow gotten into the mulberry leaves.

But Louis Pasteur showed that dying silkworms were loaded with a particular microbe he could actually see in his microscope. Isolating this microbe and injecting it into a healthy silkworm would cause the same lethal, rapidly spreading disease. This was the birth of the **germ theory of disease**, which was pursued vigorously by Pasteur in France and Robert Koch in Germany. Pasteur showed a short time later that the same thing was true for another disease—anthrax—in sheep.

Many eminent authorities of the day had based their entire lives' work on other theories of human disease, and whatever they may have thought about silkworms and sheep, they fiercely resisted the outlandish idea that human disease could be caused by tiny wiggly things visible only in a powerful microscope. But in 1891, Koch finally provided powerful evidence that tuberculosis in humans was in fact caused by a lowly bacterium. Within a very short time microbes associated with a wide range of diseases in animals and humans were discovered, including deadly diseases like tetanus and diphtheria.

Acceptance of this new way of thinking about disease was every bit as unsettling to most humans as the demonstration by Copernicus, a couple of centuries earlier, that the earth is not the center of the universe. And it must have generated a certain amount of paranoia. How do you protect yourself against something you can't see? Were people supposed to go around with a microscope, examining the skin and clothes of everyone they met to be sure they weren't carrying some deadly disease?

Again it was Robert Koch to the rescue. It was already known by the 1890s that one way bacteria cause disease is by releasing a deadly substance called a bacterial **toxin**. Students working in Koch's laboratory examined rabbits into which they had intro-

duced small amounts of toxin isolated from the bacterium causing tetanus. They gave the rabbits sublethal injections—enough to
make them sick, but not kill them. Then they looked at the **serum**
from these rabbits—the clear, straw-colored fluid left over after
blood has clotted. What they found was that the serum contained
something that was able to kill the bacteria and neutralize their
toxins in a laboratory dish.

But Koch's group wasn't finished. They injected some of this
serum, taken from the rabbit that had recovered from the tetanus
toxin, into a rabbit that had never seen the toxin before. Then they
gave the treated rabbit a shot of tetanus toxin that would have
killed any normal rabbit within a day or two. The rabbit never even
got the sniffles! They didn't know it yet, but they had just discovered **antibodies**.

The implications of these experiments were lost on no one—certainly not Robert Koch. On Christmas day in Berlin in 1891, a little
girl lay dying of diphtheria in the Bergmann Klinik. Diphtheria was
one of several infectious diseases that routinely decimated children
in many large cities at the time. Koch's group had just completed a
series of experiments that convinced them that not only could they
prevent a microbial disease with his serum transfers, but also that
they could actually *reverse the course* of a disease once it had set in.
The little girl with diphtheria was the first human being to be given
antiserum (Table 2.1). She lived, and one of the scientists in Koch's
lab most responsible for this miracle—Emil von Behring—would
later receive the very first Nobel Prize in Medicine. This method of
immunizing someone by giving them preformed antibodies, which
we refer to as passive immunization (because they themselves do
not make the antibody), is still used today in emergency situations.
It may, for example, be the only hope in case of certain bioterrorist
attacks (chapter 14) or after a deadly snake bite.

But all this was only the beginning. A new field—immunology—
had been founded, and it would yield another 22 Nobel Prizes
as researchers struggled to understand how the immune system
works.

TABLE 2.1
Some Useful Antibody-Related Terms

Antibody	A protein produced in response to invasion of the body by a microbe or other foreign biological entity
Antigen	The foreign biological entity that elicits production of an antibody and with which the antibody reacts
Serum	The fluid portion of blood remaining after the red blood cells have been clotted
Antiserum	Serum from an immune animal or person, containing antibodies
Gamma globulins	The subset of serum proteins containing antibodies
Antigenic epitope	That portion of a large, complex antigen with which an antibody specifically interacts

WHAT ARE ANTIBODIES?

The first question the new breed of scientists called immunologists asked was: what is this protective substance found in the blood of animals, exposed either naturally (through infection) or artificially (through vaccination) to microbes and their component molecules? It took nearly 30 years just to figure out that antibodies belong to a class of blood proteins called **gamma globulins**, a class that has 30 or more different kinds of molecules.

It wasn't until the 1960s that Gerald Edelman, at Rockefeller University, and Rodney Porter, working separately in England, were able to deduce exactly what an antibody molecule must look like—and in the process, earn themselves one of those 22 additional Nobel Prizes (1972). The structure they came up with consists of four protein chains, and looks essentially like the letter Y (Figure 2.1). Two of the protein chains in each antibody are long, and are called **heavy (H) chains**. Two chains in each antibody are half the size of an H chain, and are called **light (L) chains**. The two H chains are identical, as are the two L chains. At the tip of each H and L

An H and L chain pair

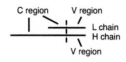

C region — V region

L chain
H chain

V region

A complete antibody molecule

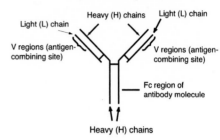

Light (L) chain   Heavy (H) chains   Light (L) chain

V regions (antigen-combining site)

V regions (antigen-combining site)

Fc region of antibody molecule

Heavy (H) chains

The five classes of antibody

IgM The antibody made when a B cell is stimulated by antigen for the first time. It is a pentameric configuration of the basic IgG structure shown below. Its pentameric nature makes it a very effective antibody (ten binding sites!), and also a very effective platform for building complement complexes. IgM is also the surface antigen receptor for new B cells.

IgA This dimeric form of antibody is found in the lining of the gut, and at other mucosal surfaces. It is highly resistant to enzymatic degradation.

IgE IgE antibodies are involved in allergic reactions, and are increase greatly during infections with protozoan parasites. It has a longer tail than the other immunoglobulins.

IgD This form of antibody found together with IgM on new B cells as the antigen, and helps with antigen recognition. It has an unusual hinge region just below the two "arms".

FIGURE 2.1
The basic structure of antibodies.

chain is a small stretch of amino acids called **variable regions,** responsible for recognizing and latching onto whatever it is the antibody is specific for. The thing that an antibody is specific for is given the generic name **antigen.**

Each antibody has two **antigen-combining sites,** formed by the variable ("V") regions of each heavy–light chain pair. These are the portions of the antibody molecule that grab onto antigen. Since the H chains in a given antibody are identical, as are the L chains, the antibody ends up with two identical antigen-combining sites. An antigen can be a bacterium, a virus, or a fungus. It can also be an individual molecule such as another protein, or a starch, or any of the complex molecules that go into making living organisms. The thing that makes an antigen an antigen, from the point of view of an animal making an antibody, is that the antigen is not found among the normal day-to-day molecules and cells of that animal (the **host**). It must be foreign. In those rare cases where antibodies form against "self" molecules and cells, the result is almost always autoimmune disease (chapter 12).

The H chains have a special region about halfway down each molecule called the **hinge region.** This region is very flexible, allowing the antibody to spread its arms apart at various angles and assume a general "Y" shape. This is also the region where the two H chains are held together by a chemical bond, as are each H and L pair a little further up. The basic antibody structure defined by Edelman and by Porter can be combined together in various ways to make different **antibody classes** (Figure 2.1), each of which has a slightly different function in the immune system, although all five classes have as their principal function the binding of antigen.

THE NUMBER OF DIFFERENT ANTIBODIES IS HOW MANY?

In order to study the antigen-combining sites of antibodies in more detail, immunologists turned to antibody-producing tumors, such as B-cell **lymphomas** or **myelomas**. All of the B cells in such a tumor produce exactly the same antibody, because all cells in a tumor are identical. They are all **clones** of the same original B cell that became cancerous. And since these cancerous B cells can be grown easily in the lab, they provide huge amounts of antibody to work with, and in particular to use for protein sequencing studies—determination of the precise amino acid sequence along the length of individual antibody chains. (Amino acids are the individual building blocks of protein molecules.)

As immunologists waded their way through ever-increasing numbers of these tumors and their antibodies, they came to a startling conclusion: there didn't seem to be any two that were alike—ever. They would occasionally come across a couple of antibodies, or maybe even groups of a dozen or more, that seemed somehow related. But two antibodies reactive with the same antigen, even when derived from the same animal, were never completely the same. There was some initial concern that tumors might somehow not reflect the "normal" world of the immune system, but it was soon generally accepted that what researchers were seeing in thousands of different B-cell tumors was in fact what was going on in the immune system itself. Each clonally distinct B cell made a completely different antibody—always.

Conservative estimates of how many different antibodies the same animal—mouse or human—could make ranged up to *a hundred million or more*. For an animal trying to protect itself from a virtually limitless number of microbes (remember the genetic/mutational advantage of microbes over mammals), that seemed like a marvelous idea. But how was all this variability in antibody structure created? Specifically, since proteins are encoded by genes in DNA, how did we come to have so many antibody genes?

Two schools of thought quickly formed. One—based on the **germline theory**—proposed that preformed genes for this huge number of different antibody molecules must all be part of the genetic inheritance passed down intact from one generation to the next. The second theory—**the somatic mutation theory**—posited that it was more likely that only a small number of "proto-antibody" genes were passed along at each generation, but those genes were somehow mutated randomly to produce an essentially unlimited number of different antibody molecules. The problem with the germline theory was that as the number of possible different antibodies an animal could make continued to grow, so did the amount of DNA needed to encode them, to a point where an uncomfortable proportion of the entire human genome would have to be dedicated to making antibodies. The weakness with the somatic mutation theory was that no one had the slightest idea how you could randomly mutate proto-antibody genes and still produce functional antibodies.

AND THE WINNER IS…

The arguments waged back and forth for a few years, but in 1976 a young scientist named Susumu Tonegawa provided definitive experimental evidence for the somatic mutation theory—and snagged yet another of those Nobel Prizes that seemed to be lying around waiting for immunologists!

What Tonegawa showed us is that in making light and heavy chains, the immune system does something no one had ever heard of or even thought of before. Proteins are normally encoded by a single gene embedded in DNA. These genes are inherited directly from generation to generation, with only minor changes brought about by mutation. They are used by each generation to guide the synthesis of the corresponding protein.

But the genes used by the immune system to produce light and heavy chains are different (Figure 2.2). Instead of a single gene encoding each H or L chain, several **gene fragments** are used. For example, an H chain is encoded by four different gene fragments.

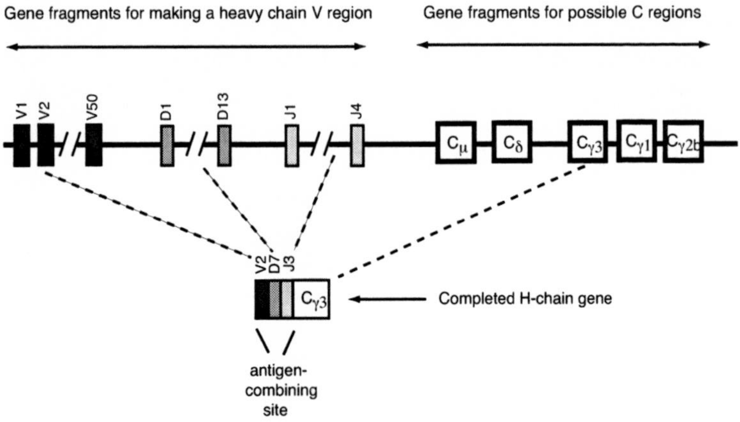

FIGURE 2.2
Assembly of a gene encoding an antibody H chain. The C-region fragment selected determines the class of the resulting antibody (Figure 2.1).

First there is a gene for the **constant (C) region**—the portion of the molecule that does not vary among different antibodies of the same class. The C region is responsible for the overall structural integrity of the antibody molecule. The V region, although much smaller than the C region, is actually encoded by three different sets of gene fragments: **V, D, and J**. Each set contains multiple numbers of fragments lined up in the DNA.

And therein lies the secret to the incredible diversity of different antibodies an animal can produce. To make an H chain, we first assemble a V region. One of the D fragments is randomly coupled to a J fragment, and then this D–J combo is stitched onto one of the V fragments to make a complete V gene. The completed V gene is then stitched onto a C gene fragment to make the final, completed gene for an antibody H chain protein.

To get some idea of how this process can generate a huge number of different antibody genes from small pools of minigene fragments, consider the following table of words. Think of the words in each set as gene fragments, and put them randomly (but sequentially) together to make sentences (V regions).

V fragments	J fragments	D fragments
Horses	eat	dogs
Fords	like	barns
Men	shoot	jeeps
Pigeons	blame	girls
Teachers	smell	mud
Liars	tickle	sardines
Children	drink	pickles

Some Resulting "Antibodies"
Teachers shoot jeeps
Pigeons tickle dogs
Children smell mud
Fords shoot pickles
Pigeons like sardines
Men blame barns
Teachers smell dogs
Liars tickle jeeps
Horses eat mud
Children blame barns
Fords tickle girls
Men smell pickles
Liars shoot sardines

Well, you get the point!

From these three groups of seven words each—21 individual words—we can make 7 × 7 × 7 = 343 different sentences, each a valid sentence, but each conveying a quite different meaning. That's exactly how we go about making antibodies that each bind a different antigen. To make an antibody H chain, for example, we humans have

V fragments   J   fragments   D fragments
50   × 6   × 27   = 8,262 different
                                                   completed H
                                                   chain genes

Each of these V genes is then attached to a C gene (which is in-variant, and doesn't contribute to the antigen-binding properties of the antibody).

A similar process in L chains can produce 433 different V regions. (Light-chain V regions are made up of only V and J fragments.)

But an antigen-combining site, remember, is formed by pairing adjacent V regions of both an L and H chain. Since L and H chains for the most part can associate randomly, humans can produce 433 × 8,262 = *3,577,446 different antibodies* just on the basis of random selection of V-region gene fragments within L and H chains, and random pairing of L and H chains.

But that's not all! During the random assembly of both L- and H-chain V-region genes, additional bits of DNA can be inserted or removed from the joining edges of V, D, and J segments. The contribution of this type of scrambling (called **junctional diversity**) to total antibody diversity is difficult to estimate precisely, but it is generally thought that humans, using their complete rep-ertoire of these kinds of genetic tricks, can make *at least 10 billion different antibodies*, and probably more—maybe 100 billion.

It would be difficult to imagine a more elegant solution of the problem posed to the immune system: devise a way to deal with a universe of pathogens (disease-causing microbes) that is not only enormous to begin with, but also in which the pathogens can ge-netically alter themselves millions of times faster than you can. The answer: bypass standard methods for shuffling the genetic deck through normal breeding processes. Come up with a whole new system for mutating genes that allows you to create not hundreds of *thousands* of new antibodies per *generation*, but hundreds of *mil-lions* of new antibody molecules per *hour*—every hour, all life long. That'll slow a pathogen down!

ARE WE SURE THIS WOULD BE ENOUGH ANTIBODY DIVERSITY TO DEAL WITH ANY POSSIBLE MICROBE, EVER?

Almost certainly. All microbes are made from the same amino acids and sugars and fats that we are; their building blocks are just arranged differently. We know from a number of studies that the portion of a protein antigen recognized by an antibody (the antigenic **epitope**) is equivalent to about six linearly arranged amino acids within a protein chain. There are only 20 different amino acids, so the maximum number of different six–amino acid epitopes that can exist in the world is 20^6, or about 64 million. The number of possible carbohydrate epitopes is in the same ballpark. So if the immune system can generate even 1 billion different kinds of antibodies, there is no way a microbe can escape detection, no matter how much it mutates itself. (Mutations don't alter amino acid or sugar building blocks, only their arrangement in protein or carbohydrate chains.)

HOW DOES ALL THIS WORK IN THE REAL WORLD?

Antibodies are made by the white blood cells called **B cells** (Figure 2.3). Each B cell is born and raised in the bone marrow. While maturing in the marrow, before it heads out to take up residence in a lymph node or the spleen, each B cell undergoes random recombination of the H and L gene fragments just described, produces the corresponding H and L chains, and assembles them into a four-chain antibody molecule.

The B cell then places a copy of this antibody on its surface, as a sort of merchant's sign saying "This is the antibody I am prepared to produce, should you need it!" The B cell uses this membrane-bound form of its particular antibody (each B cell has a different one) as an **antigen receptor**; essentially, the eyes through which it surveys the antigenic universe.

This "virgin" B cell—a B cell that has never encountered antigen—then takes up residence in a lymph node or in the spleen,

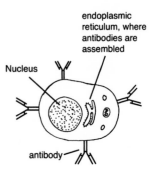

FIGURE 2.3
The B lymphocyte (B cell) displays on its surface copies of the antibody it is prepared to make and secretes millions of copies of this antibody once activated by antigen and supplied with T-cell help.

waiting for an antigen to come by that it can recognize. Now, at the time it is born, each B cell is implanted with a little self-destruct device, such that if it does not encounter an antigen it recognizes within about a week, it will die and disappear from the circulation, to be replaced by one of the literally billions of new B cells produced by the bone marrow each day.

But if a B cell runs into an antigen it recognizes via its surface antibody molecule—something that interacts chemically with its antigen-binding site—then it becomes activated and gets ready to produce its antibody and spill it out in the bloodstream, ready to go on the hunt for other copies of the antigen. In order to make this final step to an antibody-producing cell, the activated B cell will need help from the other type of lymphocyte we met in chapter 1—a T cell. We will look at the details of that step in a later chapter. For now let's just focus on what happens to a B cell after it recognizes, and has been activated by, an antigen and receives the requisite T-cell help.

This step is at the center of one of the guiding paradigms underlying the vertebrate immune system: the **Clonal Selection Theory**. Formulation of this theory was the basis for (you got it!) another Nobel Prize—this time to MacFarlane Burnet of Australia in 1960. It posits that the immune system blindly turns out billions of new B cells each day—"blind" in the sense that each cell randomly generates its antigen receptor (the antibody it is capable of producing) without any reference to the outside world. Individual receptors are not made to meet a particular threat; rather, huge numbers of them are made randomly in the hope that one or more of them may prove useful. The vast majority never do, and they (and the B cells making them) just disappear.

When one of these B cells finds an antigen, and gets T-cell help, it immediately begins to produce daughter cells—even before it starts to make antibody. And these daughter cells in turn make more daughter cells. All of these offspring of the originally stimulated B cells are clonal copies of the original, and they all make exactly the same antibody. The number of copies of the originally stimulated B cell within the total B-cell population may increase several thousandfold as a result of this clonal selection and expansion.

This is the cellular basis for what Thucydides described in the fifth century B.C.—the basis for immunological **memory**. As a result of this clonal expansion, the next time the same antigen comes into the system, instead of maybe one in a hundred billion B cells lying in wait for it, there might be one in a million. Moreover, the clonal progeny of that original B cell have been changed in subtle ways that allow them to crank up antibody production much more rapidly than a "virgin" B cell. So once you have been infected by a particular pathogen and have responded to it, the next time that pathogen enters your body it will encounter thousands of times more B cells waiting for it, each of which has a hair-trigger responding time.

This type of memory-generating immunity is also called **adaptive immunity**. Through the process of random generation of combining sites and selection of needed sites by antigen (with unused sites disappearing within a few days), the immune system is able to adapt, and constantly change in response to, the antigenic universe around it. Adaptive immunity is *the* hallmark of the vertebrate immune system, and is the exclusive property of T and B lymphocytes. Adaptive immunity is made possible by the incredible **diversity** of antigen-combining sites on both T and B cells.

MONOCLONAL ANTIBODIES

Before we leave, let's take a brief look at an extraordinarily useful research and clinical tool based on antibodies—**monoclonal antibodies (mAbs)**. Naturally occurring mAbs are the antibody products of cancerous B cells—like the lymphoma or myeloma cells we talked about earlier. Cancer cells are always clonal, so the antibodies they produce are also clonal. The mAbs they produce were useful for amino acid sequencing studies and provided the important information that the antigen-combining sites of antibodies are incredibly diverse. But these antibodies weren't useful for functional studies because it's impossible to know what antigen such an antibody is specific for. Sure, you could take one and use it to scan the entire antigenic universe until you found something it recognized, but by then you'd likely be too old to make use of it.

But there's a clever way around this dilemma, worked out back in 1975 by Georges Köhler and César Milstein (one hates to be repetitive, but—another Nobel prize!). They immunized a mouse with sheep red blood cells (SRBCs), which of course stimulated all the B cells in that mouse that recognize sheep red blood cell proteins. They then isolated the B cells from the spleen of the mouse and fused them all with mouse myeloma cells to produce **hybridomas**—hybrid tumors with the continuous growth property of the cancerous myeloma cells and the anti-SRBC antibody property of the mouse spleen cells. The hybridomas were screened to find the most specific and potent antibodies. The best of these were propagated in long-term lines that could be grown ad infinitum and would spill out endless quantities of specific anti-SRBC antibodies.

This procedure has now been adapted to a wide range of both mouse and human antibodies, which have proved to be an indispensable tool both in the laboratory and in the clinic. Because they are so pure, plentiful, and absolutely specific, they are used to establish precisely the identity of both cells and molecules. They can be used clinically in place of antisera for the passive transfer of immunity, à la Robert Koch. mAbs specific for a tumor antigen can be tagged with a radioactive marker and used to trace the location of tumor metastases within the body. It may even be possible to tag such mAbs with antitumor drugs, for direct and specific delivery to the tumor. Numerous clinical trials to do just this are under way.

Well, now you know how we have come to have an immune system that allows us to keep up with our rapidly mutating mini-assassin friends. But you don't know yet exactly how these antibodies *do it*—how they seek out and destroy or eliminate all the things that get into your body that shouldn't be there. That is the subject of chapter 3.

3

How Do Antibodies Work?

So we now know that one of the things that happens when a foreign biological substance (antigen) is injected into humans and other mammals is the production of antibodies. The modifier "biological" is important: the immune system does not waste time and energy making antibodies to nonbiological materials. Our bodies have learned during millions of years of evolution that nonbiological substances are rarely harmful. We generally rely on our livers and kidneys to get rid of them. The real threat comes from other living things—the microbes that manage to live and reproduce in us, and the biologically active molecules these parasites produce and release into our bodies. Those are the antigens that bring the immune system to a full state of alert. Antibodies are expensive to make in terms of biological energy; we don't want to waste them against meaningless threats.

Antibodies are produced by B cells, and they appear in the bloodstream about three days after the first encounter with a given antigen. If we have encountered that antigen before, and have memory B cells for it, we may see antibody in a day or two. Once made, the antibodies circulate throughout our system in search of the antigen that triggered their formation in the first place. When they find the antigen, they bind tightly to it, which triggers a series of events leading to removal of the antigen from the body.

Antibodies are extremely specific, binding only to the antigen that induced their production in the first place. It is a natural consequence of clonal selection, discussed in the previous chapter. This property of **specificity**, like diversity and memory, is a key feature of the vertebrate immune system. An antibody against the smallpox virus, for example, does not react with and lead to the elimination of diphtheria toxin, or vice versa.

The specificity of antibodies is dictated by the amino acid sequence in the V regions of H and L chains. The classic model for depicting this is the so-called **lock and key** model (Figure 3.1). The sequence of amino acids in proteins confers on them their individual three-dimensional shapes. It also makes some regions of the protein slightly greasy (lipophilic) and other areas that are electrically charged ("+" or "−"). So when an antibody in blood or lymph bumps into a protein or other biological molecule (either floating around freely or sticking out of the surface of a microbe), it will bind to it if (and only if) certain conditions are met:

1. The physical shape of the antigen must fit into the general shape of the antibody's antigen-combining site (the paired H- and L-chain V regions).
2. The electrical charges of the combining site and the antigen must attract each other; that is, positive charges on the combining site must lie opposite negative charges on the antigen, and vice versa.
3. The lipophilic regions of the antigen must be able to interact with the lipophilic portions of the antigen-combining site.

The requisite structural diversity among antigen-combining sites, to deal with the enormous range of different microbial antigenic structures, is what is generated by the gene-scrambling mechanisms we saw in chapter 2. This is what enables the adaptive immune system to discriminate with enormous power between even closely related antigenic molecules.

So antibody can recognize antigen very precisely and bind tightly to it. How does that lead to removal of antigen from the

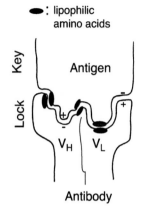

FIGURE 3.1
An antigen-combining site, formed by a VH and VL region of adjoining H and L chains, interacts with molecules it can accommodate on the basis of size and shape. This association is stabilized by mutual lipophilic and electrostatic interaction between the two molecules.

system? Interestingly, antibody by itself cannot remove antigen. When immunologists began to purify antibodies away from other components of serum, they were stumped. Antiserum containing microbe-specific antibodies could kill the corresponding microbes in a test tube. But purified antibodies *by themselves* could not kill the microbes. They would cause them to clump (**agglutinate**), but they couldn't kill them (Figure 3.2). And when purified antibodies were allowed to combine with free-swimming molecular forms of antigen, such as a bacterial toxin, the antibodies could bind to the toxin, rendering it inactive, but they didn't get rid of the toxin.

It looked as though all antibody could do is build up clumpy messes of antigen–antibody complexes. What happens to all this clumped-up gunk? Over a lifetime, you could imagine this stuff clogging up your drainpipes! But that doesn't happen; antigen disappears from the system rather quickly. So what exactly is antibody doing? To understand this, we must take a look at two more elements of the vertebrate immune system: **complement** and **macrophages**.

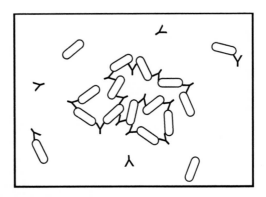

FIGURE 3.2
Antibodies clump (agglutinate) bacteria because each antibody, having two antigen-combining sites, can cross link two bacteria. A suspension of bacteria and antibodies that recognize them will quickly form large clumps that settle to the bottom of the tube, but the bacteria won't be killed unless complement is added.

COMPLEMENT: BULLETS FOR THE ANTIBODY GUN

The "clumping but no killing" conundrum was first described at the end of the nineteenth century. It was resolved through a close look at another, seemingly unrelated property of antibodies. If an antiserum capable of killing microbes in a test tube was heated to 55°C or higher, the antiserum lost its ability to kill microbes. Initially it was assumed the antibodies contained in the antiserum were destroyed, or at least their structure was sufficiently altered by heat that they could no longer recognize antigen. Loss of function in proteins as a result of heating was not unknown. So it was thought that antibodies might just be a particularly heat-sensitive protein.

But fairly quickly someone noticed that these heat-treated antibodies, even though they couldn't kill microbes, were as able to cause clumping of microbes as untreated antiserum. Clumping could only happen if the antibodies were actually grabbing on to the microbes. And that meant the structure of the antigen-combining site of heat-treated antibodies was not affected by the heat.

The mystery was cleared up when a French scientist named Jules Bordet did a key experiment. He mixed heat-treated antiserum with microbes in a test tube and saw, as others had, clumping of the microbes but no killing. When he added fresh, nonimmune serum to another tube of microbes, he saw no clumping. This was expected, because this serum contained no antibodies. But when he added some of this same fresh serum to the first tube with the clumped microbes, within minutes he saw that the microbes were killed!

This was completely unexpected, and was repeated many times to be sure it was true. Bordet concluded that there must be something naturally present in all serum, *independent of the presence of antibodies*, that somehow helped antibodies kill the microbes. This substance quickly became known as **complement**, because it *complemented* the action of antibodies in killing microbes.

It took nearly 50 years to work out what complement is and how it helps antibodies kill microbial cells. It turned out that complement is an enormously complex and delicate system. It consists of not one but over a dozen different proteins. The basic scheme for complement-mediated killing of cells is shown in Figure 3.3.

In the first step, antibodies whose combining sites recognize antigenic molecules on the surface of a cell bind to that cell, with their C-region tails (called **Fc tails**) sticking out away from the cell. Virtually every structure at any cell surface is there in multiple, usually hundreds, of copies, and so the cell will bind multiple antibody molecules. We can think of this as sort of a "tagging" process: substances that have multiple antibody Fc tails sticking out from their surface are essentially saying "get rid of me."

When two antibody molecules bind close enough together on the same cell, their protruding Fc tails serve as an anchoring site

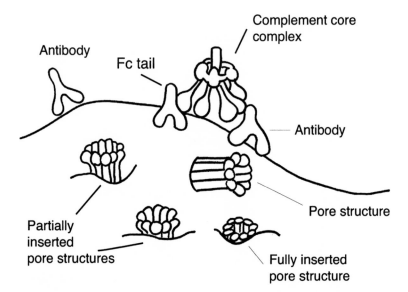

Antibody

Fc tail

Complement core complex

Antibody

Pore structure

Partially inserted pore structures

Fully inserted pore structure

FIGURE **3.3**
Some complement components bind to a cell through interaction with antibody molecules that had previously recognized something on their surface. Other complement components attach to this initiating complex and form pore structures, which sink into the membrane and allow water to enter the cell, killing it.

for the assembly of a complement complex at that very spot on the cell. The various components necessary to make this structure are present at all times in serum, and in high concentrations, so it doesn't take long to assemble a full complex. When the complex is completed, it punches a hole in the cell, and the cell dies from osmotic rupture.

Complement is part of the evolutionarily older innate immune system that we inherited from our earliest ancestors. It was around billions of years before antibodies were, and in many cases it can also recognize many microbes directly, without the aid of antibodies, although this direct process is less efficient than the one mediated by antibodies. The complement components are the same in both cases, and so is the end result—a big hole in the membrane of the microbe, followed by a quick death.

AND IF COMPLEMENT DOESN'T GET YOU . . .

Some microbes have evolved defenses against complement and can't be killed by it under any circumstances. But there is a second way of getting rid of microbes unlucky enough to have been tagged by antibody. Enter the **macrophage**. Macrophage derives from the Greek words for "big eater," and that perfectly describes what these cells do. Macrophages are huge cells—when fully activated they have a volume at least 10 times that of a lymphocyte.

Macrophages are also part of the evolutionarily older innate components of our immune system. Virtually all multicellular animals have various macrophage-like cells that roam around the body and gobble up other cells that are not part of the host organism.

Macrophages will pretty much eat anything they bump into, including microbes. But eating by casual encounter is relatively inefficient. By itself it would at best be a modest defense against infection. Fortunately, there is another way macrophages can eat that's hundreds of times more efficient. Anytime a macrophage encounters a microbe that has antibody tails sticking out from its surface, it can use a special receptor on its own surface—called an **Fc receptor**—to firmly grab onto that tail and swallow the microbe. It can do this dozens, maybe even hundreds of times before it gets full. This facilitated eating process is called **phagocytosis**.

Macrophages can also ingest free-floating antigen molecules like toxins that have been tagged with antibodies. That's why we don't see an accumulation in the body of toxin or other free antigen molecules bound to antibodies.

Once antibody-tagged microbes, or tagged molecules of free-floating antigen, are picked up by a macrophage, they are shunted into special structures within the macrophage called lysosomes. These are like little garbage disposal units. They are full of a lethal broth of poisons and digestive enzymes that rapidly degrade anything of biological origin into its component subunits (amino acids or sugars or fats). The macrophage uses what it needs to feed itself, and releases the rest for use as food by other cells in the vicinity. Waste not, want not!

The antibody response produced by B cells in response to antigen is an example of adaptive immunity. The evolution of adaptive immunity was driven, as we have seen, by the need to compete genetically with rapidly reproducing microbes. It came into existence evolutionarily only with the appearance of vertebrates. The distinguishing feature of adaptive immunity is that the response *is* modified by interaction with antigen. True, the basic genetic components of adaptive immunity are handed down relatively unchanged in the DNA that is passed from generation to generation. But the antibody molecules that result are randomly generated from the information in this DNA, as we saw in the last chapter. And most importantly, the products of this random generation are definitely shaped by the antigenic universe. Only B cells bearing antigen receptors (surface-bound copies of the antibody they are prepared to produce) that encounter their specific antigen in their environment are selected to survive and make antibody, and eventually mature to memory B cells. The composition of our memory pool acts as a record of the antigens we have encountered in our lifetime.

Innate and adaptive immunity should not be thought of as two separate immune systems. They are part of a single, coordinated response we mount against microbial infection or any other foreign biological material in our bodies—an organ transplant, for example. They work together, and in fact, as we will see in chapter 5 when we discuss the immune response to infectious disease, the function of the adaptive immune response is largely to focus and intensify the innate immune mechanisms we inherited from our evolutionary forebears.

But B cells and antibody are only part of the story of adaptive immunity. To complete our understanding of how adaptive immunity works and how it interfaces with innate immunity in keeping us alive, we must take a close look at the second major type of lymphocyte: the T cell.

4

T Cells
THE SECOND ARM OF ADAPATIVE IMMUNITY

For the first half of the twentieth century, everyone assumed that immune responses mounted by humans and other mammals must be mediated by antibodies. There were occasional hints that antibodies might not explain everything scientists saw in the laboratory or the clinic, but that in turn was assumed to mean that we didn't yet understand everything there was to understand about antibodies.

By the 1950s the number of exceptions to the rule was getting too large to ignore. For example, the rejection of transplanted tissues and organs was clearly immunological: it was characterized by both exquisite antigenic specificity and a fast, powerful memory response. Graft rejection was also accompanied by the production of graft-specific antibodies.

But these antibodies seemed to provide no protection against a subsequent transplant. When a mouse is grafted with skin from another, genetically unrelated mouse, rejection is complete in 11 to 13 days. If the mouse that had rejected the graft is subsequently grafted with skin from the same donor source, rejection takes five to seven days. "Looks like immunological memory to me," everyone said. "Must be antibodies," everyone thought.

If rejection was indeed caused by antibodies, it was reasoned, then transferring serum from the mouse that had just rejected its graft to a second mouse that had never seen that graft should result in accelerated graft rejection in the second mouse. But this was never seen. Rejection always took 11 to 13 days in the second animal, regardless of how much antiserum was transferred. The implication was that although antibodies were produced during transplant rejection, they didn't seem to play much of a role in rejection. This was a major problem in the field of transplantation immunology for many years, and made many scientists reluctant to admit that transplant rejection was truly immunological in nature.

No one was ready to go on the hunt for another mechanism of immunity, because for one thing, they wouldn't have had the slightest idea what they were looking for. But evidence that such a mechanism did in fact exist began to accumulate from a variety of sources. One of the strangest came from the study of antibodies not in humans, or even a mouse, but in chickens!

Bruce Glick was a graduate student at the University of Ohio who had become interested in a small appendix-like sac found at the tail end of the digestive tract in chickens and other birds, called the *bursa of Fabricius*. In anatomy, if the function of a structure is unknown, it is usually just given the name of its discoverer. This particular structure must have been a real puzzle—it was first described by Hieronymus Fabricius in the sixteenth century, and was never renamed! Glick tried the time-honored approach of simply removing the bursa from chickens of various ages and waiting to see what would happen. After a variety of experiments of this type, he could find no obvious differences in chickens with or without a bursa. He finally gave up and, since there didn't seem to be anything wrong with his bursa-less chickens, he returned them to the general stock.

Enter another graduate student, Tony Chang, a teaching assistant in need of a few chickens to demonstrate the production of antibodies. To save money, Chang used Bruce Glick's bursectomized chickens for his demonstration. To his immense chagrin, the chickens without a bursa failed completely as adults to produce antibody in response to antigen.

Now, at this point, many graduate students would have just shrugged and eaten the chickens. They are used to experimental failures, especially in their early years, and they're almost always hungry. But these two young men put their heads together and saw beyond the possibilities of a free meal. Together with a colleague, they carried out additional experiments that showed for the first time the important role played by the bursa in the development of the ability to make antibody. Together they wrote up what was destined to be a landmark paper in immunology, but the world wasn't ready for it quite yet. It was submitted to the prestigious journal *Science*, whose editorial staff rejected it as "uninteresting." It was finally accepted in the journal *Poultry Science*, where, as may be imagined, it languished for some years before being discovered by mainstream immunologists and suddenly became one of the most quoted papers in the field.

The most critical part of their study was the finding that while bursectomized chickens could not make antibodies, they had a perfectly normal ability to overcome many viral infections and to reject skin grafts. This was a stunning finding. By identifying and removing a specific organ controlling only antibody production, they were able for the first time to disable this powerful immune component and look at what was left over. The results showed clearly that antibodies must be only one way the immune system has of dealing with foreign antigens: in the absence of antibodies, grafts could still be rejected! These experiments forced people to begin searching for alternative immune mechanisms to explain things like viral control and graft rejection.

Other research data, as well as clinical observations, also contributed to the sense that there must be a second immune mechanism. It was found that in mice, if the thymus was removed around the time of birth (**neonatal thymectomy**), the antibody response to many bacteria was unimpaired, but the mice were unable to control most viral infections and could not reject tissue or organ transplants—the opposite of the effect uncovered by Glick's bursectomy experiments in chickens.

Meanwhile, in the clinic, the realization had begun to set in that there were two distinct categories of **primary immune deficiency**

diseases (chapter 9) arising in human infants. In one group of such diseases, typified by **Bruton's agammaglobulinemia,** afflicted individuals cannot make antibodies but are able to control many viruses and can reject skin grafts. They look, at least immunologically, like bursectomized chickens! In another category of human immune deficiency diseases, of which the most noted example is the **DiGeorge syndrome,** antibody responses to most bacteria are unimpaired, but affected individuals have difficulty controlling many viral infections and cannot reject skin grafts, just like neonatally thymectomized mice.

What all of this led to was the idea that the vertebrate immune system must have two quite distinct arms. One arm, controlled by B cells (the "B" originally stood for "bursa"), is responsible for antibody production. The other arm of the immune system was a little more complex (well, actually, a lot more complex) and was controlled by the thymus. In recognition of this fact, the cells presumed responsible for the thymus's effect on immune responsiveness were called T cells. Humans, by the way, don't have a bursa. Whatever it is that the bursa does for chickens has been taken over by the bone marrow in mammals, which works out just fine since we can keep the "B" in B cells.

THE WORLD OF T CELLS

No sooner had T and B cells been differentiated from each other than T cells had to be split into two major subsets: T-helper cells and T-killer cells (Figure 4.1). T-helper cells display on their surface numerous copies of a single protein chain called **CD4;** T-killer cells display the dimeric **CD8** molecule. Immunologists now routinely use these two terms—CD4 and CD8—to refer to helper and killer T cells, respectively. Both T-cell subsets have a receptor for antigen that is similar to but different from antibody (we'll talk about that receptor in just a moment). The pre-T cell that arrives in the thymus from the bone marrow is not yet committed to be-

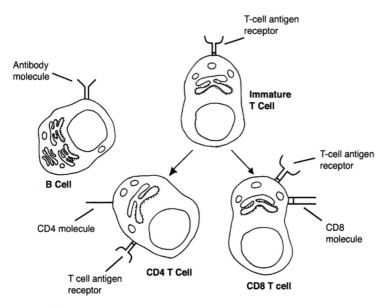

FIGURE 4.1
T and B cells, derived from bone marrow stem cells, are responsible for mediating adaptive immunity.

coming either a CD4 or CD8 cell. This happens during thymic maturation and the learning of self/not-self by a process that is still not completely understood.

An overview of the way these T-cell subsets interact with B cells and with each other is shown in Figure 4.2. CD4 cells are particularly crucial to functioning of the adaptive immune system. Someone who lacks CD4 cells cannot make many antibodies and cannot generate fully mature killer cells. They in effect have little or no adaptive immunity. CD4 cells are sometimes called "the conductor of the immunological orchestra." We'll discuss various aspects of this scheme in more detail as we proceed through this book, referring back to this figure. For now, let's just get a general sense of how the system works.

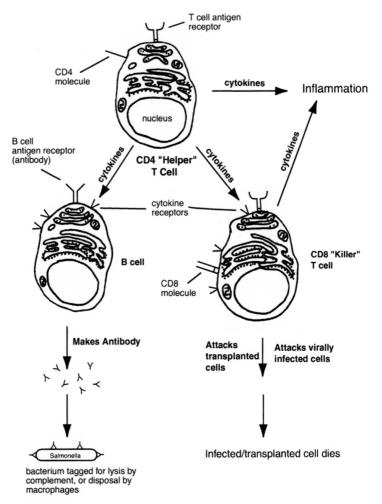

FIGURE 4.2.

CD4 helper T cells are the conductors of the immunological orchestra.

Helper T cells help two types of lymphocytes. As we said before, when B cells encounter antigen, they are not completely activated by it. They are only partially activated, and in order to proceed to full activation and antibody production they require help from a T cell. We'll discuss the nature of this help in a moment. Once they have this help, they can complete their maturation and synthesize huge amounts of antibody, which they release into the lymphatic drainage system and, via this route, eventually into the bloodstream.

T-helper cells are also required for the functional maturation of another type of T cell, the killer T cell. Killer cells were not discovered until 1960. A young physician studying kidney transplant rejection in dogs found that after a dog had rejected a kidney taken from a genetically unrelated donor, lymphocytes isolated from the transplant recipient could kill cells taken from the donor dog and grown in the laboratory. This was a completely new activity for lymphocytes and was immediately recognized as a plausible alternative to antibody as a mechanism for transplant rejection. We'll talk more about how killer cells actually destroy "foreign" cells in the body in chapters 5 and 13.

Immunologists now thought they understood how graft rejection happened—by direct, T-cell–mediated killing of graft cells. But they were puzzled why the body would be set up to defend itself against transplanted organs. This is not exactly a major threat in the life of most animals in nature! It took quite a few years to figure out that the function of killer cells in the real world is mainly in defense against viral infection. Viruses penetrate inside individual cells in our bodies and change them in ways we will discuss in a moment. Killer T cells, by means we'll also discuss shortly, can detect when a cell has been invaded by a virus. The infected cell looks different: "altered self" is the term immunologists commonly use. The immune system evolved in a world where getting rid of cells that look different from normal self was a matter of life and death for every single person, starting from the very minute of birth. To a killer T cell, cells from another person implanted in your body look sort of like your cells, but not *really* like your cells. They look different. They could, for all the killer cell knows, be your cells, but be virally infected or maybe cancerous. And they know exactly what to do in such cases.

For human beings, a lack of CD4 cells is invariably fatal without intervention, and in many cases, in spite of intervention. For example, CD4 cells are the target for infection by the AIDS virus, **HIV**. CD4 cells that have taken up HIV are ultimately destroyed, as we will see in chapter 8. There is another disease, in newborn children, called **SCID**—severe combined immune deficiency—in which a genetic mutation has rendered CD4 cells nonfunctional. These infants are, in effect, tiny AIDS patients. We will talk about SCID, and other immune deficiency diseases, in chapter 9. The lethality of these disorders is a clear indication of how dependent we humans have become on adaptive immunity.

HOW T CELLS SEE ANTIGEN

B cells "see" antigen through a cell-surface copy of the antibody each B cell is prepared to make. Almost immediately after the discovery of T cells, scientists began looking for the T-cell receptor for antigen. T cells are every bit as specific for antigen as B cells, implying the presence of an antigen receptor at least functionally like antibody. In fact, initially it was assumed that the T-cell receptor must be some form of antibody. But despite a desperately thorough search, no one was ever able to show the presence of antibody molecules in T cells. In fact, we now know the genes for generating antibody H and L chains aren't even active in T cells.

The T-cell receptor for antigen for both CD4 and CD8 cells was not discovered until 1984. It is remarkably similar to the antigen receptor on B cells. As in B cells, the T-cell antigen receptor is generated by random recombination of small genetic elements that collectively encode the complete receptor. These elements are similar to, yet distinct from, the B-cell receptor family; they lie on a completely different chromosome. But like B cells, T cells have upwards of a hundred million, possibly even a billion, possible different receptors. Each T cell also displays only one kind of receptor. T cells are selected by encounter with antigen to survive and multiply their own kind and go on to become memory cells. Unselected T cells, like unselected B cells, die out within a few days of their generation.

But even before the T-cell receptor was finally characterized, everyone knew there was going to be something very different about it. For one thing, despite repeated attempts, no one could ever show that T cells could actually bind to the antigen they were presumably specific for! For example, when an animal is immunized with a particular antigen, antigen-specific B cells and T-helper cells are both induced. If the B cells are collected after such an immunization and incubated with antigen that is either fluorescent or radioactive, it is easy to show rapid and specific binding of antigen molecules to the B-cell surface. But no one was ever able to show this kind of antigen binding to T cells. This was a real problem! How can T cells be activated by antigen if they don't bind to it? And how can we explain T-cell antigen specificity in the absence of antigen binding?

This conundrum took many years to unravel. The solution was surprising, but surprisingly simple, and it led to an entirely new appreciation of the role of T cells in adaptive immunity. It turns out that T cells do not—they cannot—interact directly with antigen molecules in solution, as B cells do. T cells *only* interact with antigen if it is complexed with a special antigen-presenting molecule called **MHC (major histocompatibility complex)**. This MHC-antigen complex is not found floating around freely in solution; it is only found on the surface of cells acting as **antigen-presenting cells**. (Chalk up another Nobel Prize: Peter Doherty and Rolf Zinkernagel, 1996!) These specialized cells are found throughout the body but are in highest concentration in lymphoid tissue and at the surfaces that line those portions of the body in contact with the environment (skin, gut, lungs).

Although several cell types can serve to present antigen to T cells, the most important by far is the **dendritic cell** (Figure 4.3). Dendritic cells look a lot like nerve cells. They have long, thin projections (dendrites) radiating out in all directions. They can be found scattered throughout virtually every tissue in the body.

Dendritic cells take up various possible protein antigens from their environment, process them, and display peptide fragments derived from them on their surface. Both CD4 and CD8 T cells constantly inspect the surface of dendritic cells, looking for anything that might be nonself. Details of what is happening inside dendritic cells as they process antigen are shown in Figure 4.4. This seems incredibly complex at first, but it is actually quite simple. It is well worth studying this figure for a moment, because it is key to understanding how T cells function inside our bodies to defend us against microbial invasion. It is also key to designing more effective vaccines for the future (chapter 7).

There are two pathways used by dendritic cells to present antigen to T cells: one for CD4 cells and one for CD8 cells. For the sake of clarity, each pathway is shown as if it existed in a separate cell, but in reality they both exist in the same dendritic cell.

In one pathway, shown on the left, materials floating around in the fluids outside the cell are taken into the cell and broken down into smaller fragments. They are taken in by a general process used by cells to take up food and oxygen called endocytosis. Substances taken in are broken down into their component parts (proteins, carbohydrates, and fats) for use as food. The proteins are broken up into smaller fragments called **peptides**. Eventually these will be broken down further to provide amino acids, a basic food source. But, uniquely in dendritic cells and a few other antigen-presenting cells, some of these peptides will be loaded onto a class II MHC protein made within the dendritic cell, and this peptide–MHC complex is shipped out to the dendritic cell surface for inspection by CD4 T-helper cells.

The second MHC pathway monitors not proteins taken in from outside the cell, but proteins that are actually manufactured *inside*

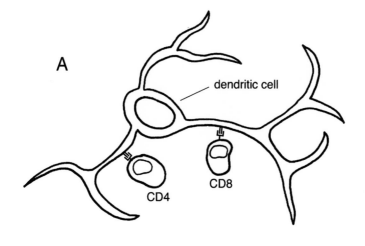

A

dendritic cell

CD8

CD4

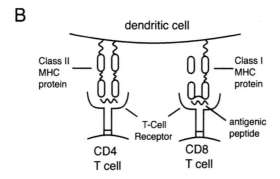

B

dendritic cell

Class II
MHC
protein

Class I
MHC
protein

T-Cell
Receptor

antigenic
peptide

CD4
T cell

CD8
T cell

FIGURE 4.3

A. Dendritic cells feed processed protein antigen to CD4 and CD8 T cells.
B. Antigenic peptides are presented on class I and class II MHC at the
dendritic cell surface and are detected by T cells using their antigen
receptors.

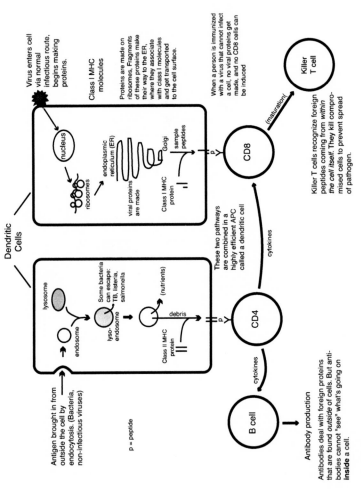

FIGURE 4.4

Dendritic cells prepare antigen for presentation to T cells. For clarity, the class I and class II pathways are presented as if existing in separate dendritic cells, but both pathways coexist in each dendritic cell.

the dendritic cell. This is an extremely important distinction. All proteins made inside any cell undergo their final assembly in a special cell compartment called the **endoplasmic reticulum**. A small proportion of the new proteins are broken down, and sample peptides are sent out to the surface of the cell coupled to a class I MHC protein for inspection by CD8 T cells. CD8 cells are intensely interested in the proteins being manufactured inside a cell, because when a cell is taken over by a virus or some other intracellular parasite, that parasite will begin making its own proteins. Some of their peptides end up at the cell surface as well, and this is a dead giveaway to CD8 T cells that something funky is going on inside the cell.

Most of the time, of course, the vast majority of materials found both outside and inside the dendritic cell are self molecules. As we saw earlier, as T cells mature in the thymus they are selected to not respond to self molecules. This learning process continues as the T cells leave the thymus and go out into the body; they are tightly regulated by a variety of means not to respond to self molecules. This is yet another very important characteristic of adaptive immunity (in addition to diversity, specificity, and memory) called self **tolerance**. A loss of tolerance to self molecules can result in **autoimmune disease** (chapter 12).

So the dendritic cell provides the T-cell component of the immune system with valuable information about what is happening inside our bodies. The class II MHC pathway that leads to CD4 T-cell activation monitors the fluids in which all cells of the body are bathed for the presence of nonself biological materials. Any such material is considered threatening, and will result in the selective activation of CD4 T cells. The CD4 cells will then go on to help B cells specific for the same antigen. (We will see how CD4 cells deliver help to B cells in the next chapter.)

The class I MHC pathway comes into play when and if the dendritic cell gets invaded by an intracellular parasite such as a virus. In the course of reproducing itself inside the cell, a virus must direct the production of all the proteins necessary to make more copies of itself. Peptides from these proteins will also make their way to the dendritic cell surface bound to a class I MHC protein, and will start the process of CD8 cell activation. Activated CD8 cells will then seek out other cells in the body displaying this same information and kill them.

But it is important to note that viruses can also get into dendritic cells through the endocytic pathway. They get scooped up from the surroundings as essentially a food source and end up being processed through the class II pathway, resulting in the activation of CD4 T-helper cells as well as CD8 killers. Both of these arms play important roles in defense against viruses.

Class I MHC molecules are present on all cells in the body except mature red blood cells. Class II MHC molecules are present only on certain cells in the body. In addition to dendritic cells, these include macrophages and B cells, two other cell types that are important for presenting antigen to CD4 T cells.

An important feature of both class I and class II MHC proteins is their tremendous **polymorphism**. All humans have both class I and II MHC proteins, but the number of different possible protein forms is extremely large (polymorphism *means* many forms). Antibodies are also polymorphic, as we have seen. But the millions of different forms of antibody are all present in each individual. The polymorphism in MHC proteins is within the *human species*, not within individuals. There are hundreds of different forms of MHC proteins, but each human will have only a few types of class I or II types present on his or her cells, and each cell will have exactly the same MHC types within the individual. MHC proteins are thus a marker of human individuality. They present a major barrier to transplantation of cells, tissues, and organs between individuals, because the possibility that any two individuals will have exactly the same sets of class I and II MHC proteins is just about zero, unless they are identical twins. We will discuss the problem of polymorphism of MHC proteins in chapter 13, in connection with organ transplantation. It was in the study of transplantation that the existence of MHC was first discovered.

T cells differ from their B-cell cousins in the consequences of activation via their antigen receptor. B cells, when activated by antigen and provided with T-cell help, produce and release into the bloodstream copies of the receptor (antibody) they display at their surface. This is not the case with T cells. When T cells are activated, they do not reproduce their receptor for release into the general circulation. Rather, they set out on patrol in the body and, using their surface antigen receptors as a set of "eyes," they themselves look for the source of the antigen.

When they find foreign antigen presented at a cell surface and are activated by it, they have a series of options for clearing the antigen out of the system. CD8 cells can kill the infected cell. Both CD4 and CD8 cells release a spectrum of small protein molecules used for cell–cell communication called **cytokines** (Table 4.1). Cytokines bind to other cells that have receptors recognizing them. Most such "target" cells are other cells of the immune system, but many cells outside the immune system (e.g., blood vessels and even the brain) have receptors for these cytokines as well. Cytokines are like the "hormones" of the immune system, in that they are substances released by one cell that influence the function and activity level of other cells. While many of them act to enhance the cells they act on, many of them, such as **tumor necrosis factor** (TNF) and the **interferons**, can cripple or destroy the cells they bind to.

Unraveling the components of adaptive immunity—T-cell and B-cell immunity—and figuring out how they work are what has driven the science of immunology since the beginning of the twentieth century. It is what even most immunologists today immediately think of when asked to define the word immunity. Humans have come to rely greatly on adaptive immunity, and usually die if it is completely absent.

But as we will see in the coming chapters, adaptive immunity *by itself* provides very little direct protection. In the struggle to survive in a world full of hostile microbes, all multicellular organisms developed innate microbial defense mechanisms that work well for them. The great advantage of adaptive immunity is not

TABLE 4.1
Representative Immune System Cytokines

Cytokine	Made by	Function
IL-11	Macrophages	Induces fever; stimulates B cells
IL-2	CD4 cells	Stimulates B cells, CD8 cells
IL-4	CD4 cells	Increases MHC II expression
IL-12	Macrophages, dendritic cells	Stimulates CD8 cells
IFN-α,β	White blood cells, fibroblasts	Inhibit viral replication
IFN γ	CD4, CD8 cells	Increases MHC I expression
TNF-α	Macrophages	Kills tumor cells; stimulates white blood cells

IFN, interferon; IL, interleukin; TNF, tumor necrosis factor.

that it supercedes innate immunity and provides us with newer and better defense mechanisms, but rather that it focuses and drives the effectiveness of innate immunity mechanisms to levels simply not seen in lower organisms.

We will keep this synergistic interaction between adaptive and innate immunity very much in mind as we begin to explore the role of the immune system—the *total* immune system—in human health and disease in the following chapters.

PART 2

THE IMMUNE SYSTEM IN HEALTH AND DISEASE

5

The Immune Response to Infectious Disease
ALL-OUT WAR!

Let's begin this chapter by quoting ourselves from chapter 1:

> You have an immune system for one reason and one reason only. In its absence, the human body would be a delightful place for microorganisms like bacteria, viruses, funguses, and parasites to live and raise their families. Your body is warm, wet, and chock full of the nutrients microbes need to survive and reproduce their own kind.

There can be no doubt that this simple truth has played *the* major role in the evolutionary shaping of the immune system we have today. The failure of any individual part of our immune defense system leads to serious disease, and often death. Infectious disease was *the* leading disease of death in humans until just over a hundred years ago. A combination of public health measures and vaccination has greatly reduced the toll in human suffering and lives, but infectious disease is still out there and can wreak enormous harm, as HIV-induced AIDS constantly reminds us. More than 150,000 people in the United States still die each year from infectious disease—35,000 from flu alone!

Each of the categories of microbes mentioned in our quote poses its own challenges to human immune defenses, and we tailor our biological response accordingly. But we spare nothing. Every single component of our immune system, both innate and adaptive, is thrown into the battle. In the rest of this chapter we will look at the immune response to the two major microbial predators plaguing humans: bacteria and viruses.

THE IMMUNE RESPONSE TO BACTERIA

For large animals like humans, who produce small numbers of offspring over long periods of time, the protection provided by innate immunity alone would be insufficient to keep the species going. But during the two to three days required to mount a primary, fully adaptive immune response to bacterial infection, innate immune components provide our only defense, and it's enough to keep us alive until T cells and antibodies kick in.

The innate immune system responds to the presence of bacteria using two major interconnected mechanisms: microbial pattern recognition and inflammation. **Microbial pattern recognition** is the recognition arm of innate immunity, the means by which all multicellular organisms detect invasion of their bodies by microbial cells. **Inflammation** is the defense mounted by the innate immune system to control microbial invasion.

Recognition of microbial invaders is based on the fact that for virtually every category of microbe, there are certain structural features that the microbe cannot change without losing its ability to survive and function in its own hostile and life-threatening environment. "Structural features" of living organisms are based on standard biochemical molecules such as proteins, carbohydrates, fats, and even DNA or RNA. When a microbe has a chemically based structure it simply cannot alter, that structure becomes a prime target for a chemically based (innate) immune defense.

Microbial pattern recognition in the innate immune system is thus based on the development, in us and our predecessors over evolutionary time, of genetically encoded proteins able to recognize and bind to those unique, unalterable microbial structural patterns *that are not found as part of anything having to do with us.* These structures are sometimes referred to as **pathogen-associated molecular patterns (PAMPs).** The receptors in our immune system that interact with PAMPs are called **pattern recognition receptors (PRRs)** (Table 5.1).

TABLE 5.1

Some Pattern Recognition Receptors and Their Corresponding PAMPs

PRR	PAMPs FOUND ON
TLR2	Bacterial cell walls; viral coat proteins; yeast cell walls
TLR3	Viral double-stranded RNA
TLR4	Bacterial cell walls; viral coat proteins; yeast cell walls
TLR5	Bacterial "tails" (flagellin)
TLR6	yeast cell walls
TLR7,8	Viral single-stranded RNA
TLR9	Bacterial, viral DNA; viral surface proteins
Nod1,2	Bacterial cell walls
MBP	Yeast, bacterial, viral surface carbohydrates

PAMP, pathogen-associated molecular patterns; PRR, pattern recognition receptors.

Functionally, the interaction of PRRs with PAMPs is no different than the interaction of an antibody with its antigen, although the defense proteins bear no particular resemblance to antibodies per se. The major difference, as we have pointed out, is that innate defense proteins are encoded in genes that are never altered in the DNA passed from generation to generation within a given species. All members of the same species have the same PRRs; every member of the same species has different antibodies. The combinatorial genetic shuffles that create T- and B-cell receptors, useful as they are, are not passed on, but must be generated anew in each individual. It is the mechanism for genetic shuffling that is passed on, not its consequences.

There are several ways PRRs work to protect us against bacteria. Some PRRs are **microbicidal**—that is, they directly kill the bacteria whose PAMPs they recognize. In humans, they are found mostly in a type of white blood cell we haven't encountered yet, called a **neu-trophil**, which is very much like a macrophage in that it is phago-cytic. **Defensins** and **cathelicidins** are two such PRRs that kill microbes much like complement: they burrow into the membrane, allowing water to leak in and cell contents to leak out. In fact, comple-ment itself can be considered an innate immune system microbicidal protein. Humans have more than a dozen variants of these micro-bicidal proteins and can use them to kill bacteria, funguses, and protozoan parasites. They can be used inside cells that have ingested microbes or released into surrounding body fluids.

Other PAMP-recognizing proteins, like the **mannan-binding lectin**, are also released into the general circulation and tag bacte-ria just like antibodies and complement fragments do. The result is the same: the tagged bacteria are eliminated by phagocytic cells that have surface receptors for this lectin. Macrophages and neu-trophils do this in humans, but almost all organisms have phago-cytic (macrophage-like) cells that carry out this function.

An important third category of innate PRRs signal white blood cells that there are bacteria in the neighborhood. These PRRs sit in the membranes of the white cells and interact with many of the same microbial PAMPs that circulating defense PRRs do. In humans, the most important of these proteins are called **Toll-like receptors (TLRs**; Table 5.1). They were originally discovered in fruit flies, where they play a role in shaping the developing fly embryo, but are also involved in the fly's microbial defenses. Flies that lack them are very susceptible to fungal infections. There are about a dozen different TLRs in people, distributed on those cells—dendritic cells, macrophages, and even B cells—that are on the front lines for fight-ing microbial invasion. When these proteins are missing or lose func-tion, humans also are more susceptible to a variety of infections.

When a white cell bearing a TLR comes into contact with a mi-crobe displaying a matching PAMP, the cell becomes activated and helps kick off an inflammatory reaction. This is the innate immune response in its fullest, most lethal glory (Figure 5.1). Inflammatory reactions have been described since ancient times, based on im-

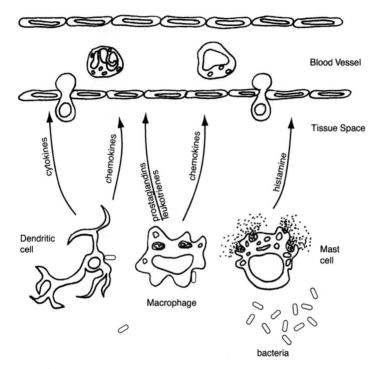

FIGURE 5.1
The inflammatory response involves changes in local blood vessels.

mune reactions to bacteria or other foreign substances taking place at the surface of the body, such as at the site of a wound. They were described by early physicians as involving redness, heat, swelling, and pain. The redness and heat derive from blood rushing into the affected site, which together with accumulating lymph also causes swelling and pain from the resulting pressure on local nerves.

Although not obvious until the past century or two, the same set of reactions so obvious in the skin can also occur anywhere in the body that foreign biological agents manage to get to. **Hepati-**

tis is inflammation of the liver, triggered by a virus; **colitis** is inflammation of the colon, usually caused by a parasite.

The most important TLR-bearing cells involved in initiating an inflammatory response are the dendritic cells and macrophages. Once they have been activated by interaction of their surface-bound TLR with a microbial PAMP, they release a spectrum of **proinflammatory cytokines** that trigger most of the features of inflammation.

One of the first things cytokines such as **prostaglandin** do during inflammation is alter nearby blood vessels. Some cytokines cause the blood vessels to enlarge, which can be seen at the surface of the body as a reddening of the skin and an increase in local heat. Bacteria also activate complement, either directly or with the aid of antibodies, which results in their lysis or phagocytosis by macrophages.

But fragments of complement released in the process also activate **mast cells.** Mast cells in turn release **histamine**, which also acts on local blood vessels. All of these agents also cause the blood vessels to become more leaky, so that white blood cells can cross out of the bloodstream and into the site of the infection (**extravasation**).

Yet other cytokines (often referred to as **chemokines**) released by macrophages and dendritic cells act as attractants to guide infiltrating white cells to the source of the infection. One of the first to arrive is the neutrophil, which also expresses TLRs, and these cells immediately pitch in to help clear bacteria from the area by phagocytosis and by release of microbicidal proteins. A short time later more macrophages also arrive, and then lymphocytes, both T and B, infiltrate the area.

The inflamed area becomes swollen with the arrival of all these cells and the accompanying fluids and may become painful as well. Some cytokines, such as **TNF-α** stimulate local pain receptors, which alerts our brains to a problem in the region. Other cytokines go directly to the brain and trigger fever, which compromises the survival of many bacteria. Painful damage can also come from the fact that both macrophages and neutrophils are sloppy eaters and dribble some of their digestive juices into the area, damaging normal cells and tissues—not pretty, but effective.

The dendritic cells that were the start of much of this innate activity also serve to jump-start adaptive immunity. Dendritic cells can ingest bacteria in a phagocytic process facilitated by interaction of dendritic cell TLRs with bacterial PAMPs. Once they have ingested bacteria or other microbes, dendritic cells move into an activated state and migrate toward a nearby lymph node. Along the way, they process microbial proteins and present the resulting peptides at their surface for examination by T cells once they reach the node. They also provide the T cells with cytokines the T cells need to become fully activated.

As we saw in Figure 4.4, the class II pathway of antigen processing results in CD4 T-helper cells ready to provide help to B cells. The activated T-helper cells break away from the dendritic cell and go on the hunt for B cells needing their help. What was not obvious in this figure is exactly how CD4 cells deliver this help.

B cells use their surface antibody molecule to find and ingest bacteria, process bacterial proteins through the class II pathway, and display bacterial peptides on their surface in association with class II MHC. During an infection by a particular strain of bacteria, the same overall patterns of peptide–MHC structures will be found on B cells and the dendritic cells. So after the CD4 cell is activated by a dendritic cell displaying foreign peptides, it seeks out B cells displaying the same information. When the CD4 cell finds a B cell it recognizes and binds to it, the CD4 cell releases its full spectrum of helper cytokines.

In certain situations, B cells can also become activated to produce antibodies without T-cell help. This occurs through a B-cell PRR interaction with a microbial PAMP. B cells activated in this way go on to produce antibody. This happens much faster than ordinary T-cell–facilitated B-cell activation. However, this kind of **T-cell–independent B-cell activation** is indiscriminate. It does not represent activation by specific antigen; any and all B cells coming into contact with these cell-wall products via their PRRs will begin to produce antibody, whether those particular antibodies are needed to fight the infection or not. This is wasteful, but in terms of getting antibody out there in the shortest possible time, it might make the difference in getting the upper hand on an infection. B cells activated in this way produce only IgM antibody and do not go on to become memory cells.

With only a few exceptions, bacteria spend their time in mammals like us living and reproducing in the bloodstream, so the main adaptive response to bacteria is the production of antibodies. As we have seen, the antibody tags the bacteria for killing by complement or engulfment by macrophages. Neutrophils also love antibody-tagged bacteria and clear them very effectively. But there are a few cases in which bacteria can actually crawl into cells and live and reproduce there. This requires a different kind of immune response.

T-CELL IMMUNITY TO BACTERIA

Tuberculosis (TB) is a disease of the lungs caused by the bacterium *Mycobacterium tuberculosis*. Like smallpox, it was once a major cause of death in humans, particularly among working people living in crowded, unhygienic urban quarters. Unlike smallpox, which usually killed its victims quickly, TB took much longer, toying with its victim for years, sometimes decades, before death finally came. As such, it was a major cause of chronic illness, as well as death, and wreaked major havoc on the economies of nineteenth-century societies. Its victims lingered on, unable to work and generate income but requiring care and sustenance from their families and friends.

TB is the first *human* disease actually shown to be caused by a microbe. In 1892 Robert Koch presented experimental proof for this claim at an international meeting of physiologists. Koch's presentation shook both the scientific and the lay worlds. The evidence he had gathered in support of his claim was irrefutable, and everyone who was at the meeting, or who read the next day's newspapers, realized they were seeing the dawn of a new era in human medicine.

Koch went on to show that injection of a protein extract of these bacteria called *tuberculin* into the skin of someone who had recovered from tuberculosis caused a transient inflammatory skin reaction. Within a day or two the skin in the area of the injection became red, itchy, swollen, and tender. The person might also develop a mild fever. In animal experiments, where larger amounts of tuberculin could be injected, the fever became substantial, and the skin reaction often became necrotic and ulcerous. These kinds of secondary skin reactions to antigen were known as **hypersensitivity reactions.**

The tuberculin reaction proved to be a bit of a mystery. Hypersensitivity reactions to other microbes and microbial products, as well as nonmicrobial antigens, were well known. But in all previously known cases, the skin reaction would develop immediately, within minutes of injection of the provoking antigen, and would usually resolve within a day. The tuberculin reaction, on the other hand, could not be detected for a number of hours after injection of antigen, peaked 36 to 48 hours later, and might not subside completely for several days after that.

The "delayed" nature of this hypersensitivity reaction was generally recognized, but was not at first considered sufficient reason to view the tuberculin response as fundamentally different from other, more immediate hypersensitivity reactions. In addition to the tuberculin reaction, this kind of **delayed-type hypersensitivity (DTH)** reaction was eventually seen in skin reactions to things like poison ivy, poison oak, and sumac.

But like graft rejection, DTH reactions *were* fundamentally different from the more rapid **immediate hypersensitivity** reactions. Immediate hypersensitivity, to substances that cause allergies, for example (chapter 10), could be transferred from one individual to another with serum. Injecting some of the provoking substance under the skin of the recipient of the allergic serum elicited the same rapid, irritating reaction. However, the delayed skin response to tuberculin could not be transferred with serum, either in animals or in humans. When, for example, serum from a tubercular guinea pig was transferred to a healthy animal, the recipient showed no skin reactivity at all when tuberculin was injected intradermally.

The inability to transfer delayed hypersensitivity with serum ruled out antibody as a causative agent. Yet DTH reactions, like immediate hypersensitivity reactions, were shown to be absolutely antigen specific and had the property of memory. If antibodies were the only agents with the property of specific antigen recognition, how then could DTH be antigen specific? This was precisely the same dilemma that faced transplant immunologists, as we saw in chapter 4.

This puzzle was finally resolved by a milestone experiment carried out in the 1940s. It was shown that antigen-specific delayed hypersensitivity could be transferred between animals using *cells* (lymphocytes) from the hypersensitive animal, rather than serum. This experiment is one of the most important in immunology, because it was the first to show that cells, as well as antibody, can have the properties of antigen recognition and memory. It led to the conclusion that the immune response to at least one bacterial disease—tuberculosis—was mediated by immune cells themselves, and not antibody.

This experiment marked the beginning of a major subdivision of immunology called *cellular immunology*. Cellular immune responses would eventually be recognized as the basis for a wide range of immunological phenomena, including (in addition to DTH reactions) transplant rejection, suppression of viral infections, many autoimmune diseases, and some aspects of tumor control. All of these reactions are now known to be mediated directly by T cells.

As we have seen, the main contribution of T cells, both CD4 and CD8, is in the production of cytokines that drive both innate and adaptive immunity. In addition, CD8 T cells can become cytotoxic; that is, they can kill self cells that are judged to be compromised.

CD8 killer cells are thought to play an important role in overcoming infections where the microbes, as in the case of tuberculosis, hide inside cells. We will discuss this aspect of TB further in the next chapter. But one of the most important roles for cellular immunity is in defense against viruses, to which we now turn.

THE IMMUNE RESPONSE TO VIRUSES

The first thing to understand about viruses is that they are not alive. Viruses lack any of the structural characteristics of a living cell. They have a coat, but there is almost nothing under it: no nucleus, no mitochondria, no ribosomes. Yet, if we allow that the single most important task of living things is to pass on their genes to as many offspring as possible, then viruses are very much a form of life—some kind of "bioid"—for the one thing viruses do have under their coats is genes.

In fact, viruses are nothing more than DNA (or RNA) wrapped in a few strands of protein, with the occasional lipid thrown in. And by the criterion of reproductive capacity, pound for pound viruses may be the most efficient biological entities on the planet. The fact that they must infect a living cell to reproduce should not be held against them. In getting someone else to do most of the work for them, they might well be viewed as among the most successful of all living things—except, of course, they're not alive.

Since viruses aren't alive, the immune system can't use defense strategies like "viricidal proteins" or complement against them; there is nothing to kill. This is the same reason antibiotics are useless against viruses. One of our most important defenses against viral infection is something called Type 1 interferon.

There are two main types of Type 1 interferon: **interferon-α** (IFN-α) and **interferon-β** (IFN-β). These cytokines are produced by many white cells and by connective tissue cells scattered throughout the body, which are not actually part of the immune system. When viruses infect any of these cells, the cells begin to produce Type 1 interferons, which can inhibit viral replication. Interferon is also produced, in much larger quantities, by dendritic cells that have been infected by a virus. The interferon is released into the general circulation and picked up by nearby cells that have not yet been infected, making them highly resistant to subsequent viral infection.

But as with bacteria, infection by viruses also provokes a rapid, vigorous immune response, involving both the innate and adaptive immune systems. Both dendritic cells and macrophages have Toll-like receptors that interact with various PAMPs on viruses. This in itself is enough to get inflammation off to a start via release of proinflammatory cytokines. We see the same alterations of nearby blood vessels, the same rapid influx of the same white cells as in a bacterial infection. And as with bacterial infection, the inflammatory response to viruses is followed fairly quickly by an adaptive immune response facilitated by dendritic cells.

Both antibodies and T cells play important roles in the adaptive response against viruses. Virus-specific antibodies can tag viral particles and prepare them for phagocytosis by macrophages or neutrophils. In fact, for many viruses, antibodies, once produced, provide an adequate defense against the infection. T cells produce their own interferon, interferon-gamma (IFN-γ). This cytokine is a potent activator of macrophages, driving them to a literal frenzy of eating activity, which helps clear the infection. It also increases the expression of class I MHC on cell surfaces.

But a special problem arises in the case of those viruses that do not spend much time in the circulation—in the blood or lymph. Many viruses grow for a while inside a cell, and then cause the cell to burst open, killing it and releasing viral particles into blood and lymph. These make their way to new cells, and as they are moving in transit through bodily fluids they are easy targets for antibodies.

But other viruses spend most of their life cycle inside a cell, without killing it. They reproduce more slowly, and release new viruses from the cell surface, without damaging the cell. The new viruses may migrate only a short distance to infect a nearby cell. In these cases, antibody can be relatively ineffective, or not effective at all.

Antibodies remain in the circulation: they *cannot cross into cells* where most of the viruses are hiding. They must wait to pick off the rare, lonely virus caught scurrying from one cell to another.

This is where we need killer T cells. As we saw in Figure 4.2, viruses can get into cells two ways: through the endocytic route, to produce a T-helper cell response via the class II pathway, or by actual infection of the cell, in which case the virus inserts its genetic information into the cell's nucleus and directs the cell to begin making viral proteins. Peptide fragments of these proteins make their way to the cell surface via the class I pathway, where they are detected by CD8 T cells.

CD8 killer cells, once activated by viral peptides presented by dendritic cell class I, kick off into the circulation and go looking for other cells in the body that are harboring the same virus and displaying the same peptide–MHC complex on their surface. When they find such a cell, they bind to it and deliver a so-called **lethal hit**. The killer cell then detaches and moves on to find other infected cells. A few moments later, the infected cell it just detached from dies. It swells to several times its size, its membrane begins to undulate faster and faster, and finally the cell breaks up into small membrane-bounded vesicles that are eaten by neighboring cells.

The nature of the lethal hit was a matter of intense study for over 20 years. How did CD8 killer cells do it? What made them such efficient killers? Everyone was looking for a knife or a gun, a rope or some poison. But no such weapon was ever found. The answer, when it was finally uncovered, was delightfully simple yet sophisticated.

It turns out that all cells in our bodies have a built-in suicide program called **apoptosis**. This program is brought into play in various situations. Sometimes, embryos in utero make cells that are needed at one developmental stage but are not needed later on, and these superfluous cells are instructed to commit suicide. For example, at about five to six weeks human embryos have what look like paddles instead of hands and feet: individual digits laced together by a delicate webbing. But on cue, the cells that make up the web between condensing fingers and toes in these duck-like paddles commit suicide, and our digits emerge, beautifully formed. Other times, a cell's DNA may become damaged, which is a risk for cancer, and if the cell cannot repair the damage, it too is induced to commit suicide. Cell suicides of this type are very common and widespread in biology. They have been around for at least a couple of billion years.

CD8 killer cells know how to turn on this program. It's as simple as that. When the killer cell recognizes that a cell is harboring a virus, the rare intracellular bacterium like *M. tuberculosis*, or any other intracellular parasite, it just tells that cell to turn on its suicide program. It has a couple of ways of doing this. It can deposit a protein called perforin, distantly related to complement, on the infected cell's surface. As you can probably guess from the name, it was originally thought (based on its relationship to complement) that perforin probably punched a whole in the infected cell's membrane, inducing it to die by osmotic lysis. That doesn't happen. The cell, by means we don't fully understand yet, translates the presence of perforin on its surface as a command to commit suicide.

There is a second way activated CD8 cells induce their targets to commit suicide. CD8 cells have a molecule on their surface called **Fas ligand.** When they decide a target cell needs to die, they just insert this "key card" into a **Fas** "lock" on the target cell surface, and this immediately activates the same apoptosis suicide program.

And finally, in some situations, cytokines released by CD8 cells, such as IFN-γ and TNF-α, can directly or indirectly compromise cells recognized by CD8 killers. We will be particularly interested in this when we look at the role of CD8 cells in fighting cancer and in rejecting organ transplants (chapters 11 and 13).

A NATURAL KILLER

In addition to CD8 killer cells, there is another cell we haven't met yet, called the **natural killer cell**, or **NK cell**, which also induces compromised cells to commit suicide. NK cells were originally discovered while looking at the body's immune response to cancer. We will look at this subject, and the role of NK cells in it, in chapter 11.

NK cells are now known to play a major role in defense against viruses, and in fact that may be a more important role for NK cells than cancer surveillance. NK cells are driven to a highly activated state by the IFN-α and -β produced early in a viral infection. NK cells are considered part of the innate immune system for two reasons. First, they do not have to be activated or sensitized: they are ready on the spot to kill a tumor cell or a virally infected cell without ever having seen one before. Second, NK cells do not have to scramble their genetic information to produce a broad repertoire of antigen-specific receptors. Everything they need to do their job is up and running from the moment of their birth.

Evolutionarily, however, NK cells are a recent addition to innate immunity. There is nothing quite like them prior to the vertebrates. This just underscores the fact that innate immunity is not a matter of evolutionary stage, but rather a mode of action. Like CD8 killer cells, NK cells use perforin, and to a lesser extent Fas, to induce apoptosis in their targets.

NK cells plug a gap created by clever viruses trying to escape the deadly CD8 killer cell. Some viruses have learned to inhibit their host cell's expression of class I MHC proteins as a way of blocking migration of their peptides out to the host cell surface. This completely blinds the CD8 killers and knocks out their involvement in controlling such viruses. In the case of viruses that hide long term inside cells, that could be disastrous (for *us*!).

Enter NK cells. Although originally thought to be involved only in the body's response to cancer, it soon became apparent that they were also effective against viruses, but only certain viruses. For many years, no one could figure out how NK cells detected the presence of a virus inside a cell. NK cells do have TLR receptors, which could account for their *activation* by viruses in the circulation. But how were they recognizing cells as being virally infected and delivering the lethal hit? They did not have any receptors for viral peptides. What were they seeing? And why only cells infected by *some* viruses? The answer, once again, was delightfully simple yet sophisticated. They were seeing nothing. Yes, nothing!

It turns out that NK cells kill any (and *only*) cells not expressing a class I molecule. A class I molecule on a cell actively inhibits an NK cell from killing it. It appears that NK cells are completely ready and able to kill any cell they meet, but if that cell displays class I, the NK cell's killing mechanisms are shut off. If a cell lacks class I—and no normal, self-respecting healthy cell would *not* express class I molecules—it is immediately induced by the NK cell to kill itself—unbelievably clever. MHC class I: don't leave home without it!

BACTERIA WE CAN LIVE WITH

You may be aware that a number of strains of bacteria live in mutual coexistence within us and are fed and nurtured by us in exchange for work they do on our behalf. Indeed, these **commensal** microbial cells, representing many hundreds of distinct bacterial species, may actually outnumber the human cells that make up our bodies! (Since they're less than a millionth of our cells in individual size, they're hard to notice.) They live mostly in the large intestine, or colon, and help us break down food that would otherwise pass through unused, and they also make and secrete a few things that we need, such as certain fats and vitamin K_1. Because they are in the gut in such high density, they make it difficult for other, potentially pathogenic (disease-causing) bacteria to gain a foothold.

How did these bacteria get there, and why doesn't our immune system kick them out? We are not born with these helpful bacteria, but must take them in from our environment. Some come in with food; many come in from kids just playing in dirt. We know from experiments in mice that failure to take in such bacteria can cause major nutritional problems, and it is thought that living in an increasingly germ-free (in relative terms) environment may also cause problems for humans including, aside from digestive difficulties, things like allergies and autoimmune disease.

But still, how does the immune system know these bacteria are "good guys"? Why doesn't it deal with them in the same way it does other microbes? Do immune cells that encounter commensal bacteria in the gut not have TLRs or other PRRs, or do the bacteria lack PAMPs? Do they not see these bacteria, or do they see them and just ignore them? The answer may be a combination of both, but the bottom line is that we don't really know.

We do know that if bacteria commensal in the gut enter into the general tissue spaces of the body, they are rapidly destroyed by the immune system. On the other hand, there are potent immune elements present in the gut itself that can act on and destroy noncommensal bacteria, largely through the antibodies of subclass IgA (Figure 2.1). This tells us two things: commensal bacteria are not immunologically invisible and can trigger immune responses outside the gut, and the immune elements in the gut have what it takes to spot potentially harmful bacteria.

There is in fact good evidence that the immune system within the gut does react to intestinal bacteria and can kill them, but mostly it keeps them from entering the general circulation, where they would be rapidly and completely destroyed. The immune system and commensal bacteria seem to have developed a delicate balance, whereby the bacteria survive sufficiently well to provide needed functions but do not expand to a point where they would threaten their host (us).

The exact basis upon which this balance is achieved is still obscure but may involve modulating PRRs on host immune elements and PAMPs on the bacteria. It seems likely that commensal bacteria may also release moderately immunosuppressive molecules of various sorts, which slow the immune response mounted against them. Trying to unravel the intricacies of the dance between our gut and its commensals has occupied immunologists for nearly a century, and it is not clear we know much more now than we did when we started.

The immune system takes microbial invasion very seriously. The mechanisms it employs are far more extensive and complex than could ever have been imagined at the dawn of the age of immunology. Still, inflammation—one of the first manifestations noted by humans of the power underlying immunity to infection—remains the ultimate weapon in our antimicrobial armamentarium. It is a violent response, and as we will see in coming chapters, the collateral damage it causes to healthy tissues can be deadly, even fatal—but it works. The fact that we are still here is mute testimony to its effectiveness.

6

When the Immune System Is the Problem, Not the Solution

MICROBIAL IMMUNOPATHOLOGY

The immune response to microbial infection is massive and brutal. It has to be. Given the rate at which unhindered microbes could reproduce in our bodies, we could be reduced to soup in no time flat. Just look at what happens to humans after they die—most of that is done by microbes.

While we are alive and fighting the microbes that would like to make soup of us, the all-out, take-no-prisoners impulse of our immune system works very well. Most infections are quashed on the spot. The weapons used to destroy microbes are potent and can do considerable harm to us as well as them. But if the attack is limited in scope and resolved quickly, damage to normal tissues is usually minimal, and the immune system secretes a number of cytokines that promote healing once the battle is over.

But we can get into trouble if our initial immune attack fails to clear an infection quickly and completely—when acute infections become chronic. Then a game of hide and seek gets under way. The microbes get the upper hand for a while, and then the immune system steps up its efforts, only to be overwhelmed again by some devious strategy of the microbes. As time passes and the infection does not resolve and move on to the healing stage, the collateral damage caused by this constant, see-saw warring begins to take a toll—and it can be fatal.

In the sections that follow, we will look at two cases where this form of **immunopathology**—disease caused not by the invading microbe, but by the immune system itself—is a particular problem: **tuberculosis**, which we looked at briefly in the last chapter, and infection by the **hepatitis B virus**. We will also look at a situation in which the immune system seems initially to overreact, but then appears to become paralyzed: **bacterial sepsis**.

TUBERCULOSIS

Tuberculosis (TB) is certainly an example of microbes provoking an all-out attack by the immune system. The bacteria that cause TB are spread from one individual to another in aerosol form. One individual releases them by coughing or sneezing. Being very light, they may stay aloft in the air for some time before settling on various surfaces, where they will usually die within a short time. But while still airborne, they may be breathed in from the surrounding air by a complete stranger and settle into a new set of lungs.

The new host is not defenseless; the **tubercle bacilli**, as they are called, are immediately engulfed by macrophages in the lungs. ("Bacilli," the plural of bacillus, is a term for a particular subtype of bacteria.) Like all macrophages, their job is to eat everything in site and release proinflammatory cytokines that kick off a full-scale inflammatory response. Depending on the infectious strength **(virulence)** of the invading bacilli and the defensive strength of the macrophages, the bacilli may be completely destroyed by the macrophages, in which case—end of story. You very likely will never know it happened.

But tubercle bacilli have developed a clever trick to survive a "big mac attack." It doesn't always work, but it works often enough to keep these microbes going as a species. The bacteria are taken up into macrophage **phagosomes**, as shown in Figure 4.4. If those phagosomes fuse with the lysosomes, the tubercle bacilli will be toast. But these bacteria produce a substance that prevents the phagosome from fusing with lysosomes. The bacilli then reproduce within the phagosome, using the macrophage's general food supply (which probably includes, by the way, some of their previously digested cousins and aunts!). This is one of the most insidious and dangerous tricks developed by pathogenic microbes. It's like having a night stalker taking up residence in your attic, crawling around through your walls and ceilings while you sleep, raiding your pantry and larder, and rearing a family while waiting for a chance to do you in. A number of other bacteria have developed this same trick. As we will see in chapter 14, one of them, *Francisella tularensis*, has been classified as a potential bioterrorism agent.

Once bacilli take up residence inside a macrophage, they will continue to increase in numbers until, at some point, the macrophage bursts open and releases thousands of fresh bacilli into the surrounding tissue. This attracts even more macrophages to the site, which promptly ingest the newly released bacilli. The disease could still stabilize at this point, if the new macrophages manage to kill off most of the bacilli released from dying macrophages.

But as the bacteria continue to reproduce inside macrophages, some of them do get digested, and fragments of bacterial proteins like tuberculin make their way to the macrophage cell surface coupled to class II MHC proteins. This draws the attention of CD4 T cells lured to the site by inflammatory cytokines. By a special trick available to dendritic cells and macrophages, some of these fragments also get into the class I MHC pathway as well, which brings CD8 cells into the game.

Both CD4 and CD8 T cells release cytokines (particularly interferon-γ that stimulate macrophages and other elements of the innate immune system to even greater efforts. This is another critical stage. In many cases, activation of T cells may be sufficient to arrest progress of the disease and clear the bacilli from the host. This is the stage where transfer of lymphocytes from infected individuals into naïve individuals can transfer a delayed-type hypersensitivity (DTH) reaction to tuberculin. The CD4 cells can also help B cells make antitubercle antibody, which is used to tag tubercle bacilli floating around outside of cells and speed up their ingestion by macrophages.

But when even this response fails, things can turn ugly. When the T cells sense that in spite of their efforts to attract and stimulate macrophages, in spite of making copious antibody, the infection persists, both CD4 and CD8 cells mount a broad-scale DTH reaction that starts to compromise lung function. CD8 killer cells begin killing the infected macrophages, depriving the bacilli of a place to replicate. They also release a molecule called **granulysin** that can kill escaping bacilli. But many of the bacilli do escape and make their way ever further into the lungs, where they continue to replicate and, less hindered now by hungry macrophages, begin to infect healthy lung tissue.

The CD8 cells then proceed blindly to kill off those infected lung cells not already destroyed by the bacilli replicating inside them. There follows the disease stage with the ominous name "liquefaction and cavitation." Large sections of lung tissue are literally melted away by disease.

But notice that it is not the bacteria themselves that are the major culprit now; in these late stages of the disease, *the vast majority of the damage is done by the host's own T cells,* in what is essentially a prolonged, chronic DTH and killer-cell attack against normal lung tissue. This deadly hide-and-seek, search-and-destroy action game between the microbes and the immune system is almost always the cause of death from tuberculosis. A similar insidious immunopathology causes the damage in **leprosy,** which is caused by the related bacterium *Mycobacterium leprae.*

As a footnote to this story, we should note that after 20 or so years of decline in both incidence and mortality, tuberculosis is once again on the rise around the world. About 3 million people worldwide will die this year from tuberculosis. The World Health Organization (WHO) has identified TB, along with malaria and AIDS, as a critical focus for new vaccine development (chapter 7). In the United States, we may see 25,000 new cases annually, with perhaps 2,000 or so dying from the disease.

Part of the explanation for the recent rise in the incidence of TB is doubtless the appearance of AIDS (chapter 8), which destroys the T-cell system and renders individuals more susceptible to diseases like tuberculosis. One in seven AIDS victims worldwide dies of TB. Other factors may also be involved. A recent analysis by the

Centers for Disease Control suggests that one of the major reasons for increased mortality in recent years (as opposed to increased incidence) has been noncompliance with physician-recommended treatment schedules for tuberculosis. And as TB-active AIDS patients are treated with ever more powerful doses of TB-fighting drugs, we are seeing the emergence, through mutation, of new extreme drug-resistant strains of *M. tuberculosis* (XDR-TB strains) that have become resistant to these drugs. In South Africa, where such strains are an acute problem, 52 of 53 patients infected with these new TB strains died over a three-month period in 2005.

Viral Hepatitis

In the case of tuberculosis, the invading tubercle bacilli certainly can't be considered harmless. If left unchecked by macrophages and T cells, they would doubtless destroy the lungs on their own. One could argue that if the pathogen is going to kill you anyway, not much is lost if the immune system kills you while trying to clear the infection. In that light, the subsequent overreaction by the T cells, while regrettable, is at least understandable.

But in the case of infection of the liver with the hepatitis B virus (HBV), it is a little harder to be so understanding. HBV-induced viral hepatitis (also known as serum hepatitis) is truly the modern equivalent of smallpox. It affects nearly 350 million people worldwide, and is today one of the world's leading causes of death from infectious disease. It spreads from person to person mostly via contact with body fluids such as saliva, blood, vaginal secretions, or semen. In developing countries it is also spread by contaminated needles used for injections of various sorts.

Like HIV, HBV spreads rapidly among male homosexuals and intravenous drug users, 80% or more of whom show evidence of exposure to the virus (compared to 5% in the general population). It induces both an acute and a chronic form of hepatitis, either of which can be fatal, and in its chronic form is also a leading cause of liver cancer. The initial symptoms are usually quite mild, little different from a mild case of the flu. It is nonetheless virtually impossible to treat once it is established. The course it takes is entirely dependent on how the immune system decides to deal with it.

Although only one of several viruses that can cause liver disease, HBV is by far the most damaging to human beings. The liver damage in HBV hepatitis can be massive and devastating. Yet, so far as we know, *HBV itself causes no harm at all to liver cells!* Outside the body, liver cells infected by HBV get along just fine; there is no sign of virus-induced damage. Some individuals become tolerant of the virus and do not react against it immunologically. Although loaded with virus, they show no signs of the damage seen in hepatitis. All evidence suggests that in this disease, when serious damage occurs, it is *the immune system* that causes most, if not all, of the damage.

In tuberculosis, remember, the pathogenic tubercle bacilli invade macrophages. In the course of trying to destroy pathogen-altered macrophages, the T cells end up destroying the lungs. A similar but even more deadly sequence of events takes place in HBV-induced hepatitis. Like all viruses, HBV invades normal cells and takes over the cell's machinery in order to make more HBV. In the process, HBV introduces its own small piece of DNA into the infected cell's nucleus. Once this happens, the cell treats the viral DNA just like its own. It copies out the HBV instructions for making more HBV, and at the same time copies out viral instructions that interfere with many of the cell's own normal functions. This is much sneakier than tubercle bacilli, which we previously likened to a prowler crawling around in the attic and raiding the pantry. Invasion with a virus like HBV is much more like someone living inside your own skin, taking over your body, and pretending to be you while using you for its own ends. It's a very clever, and potentially very deadly, strategy.

But as we have seen, each cell in the body displays on its surface samples of the proteins it is currently making. A virally infected cell sends viral proteins out to the surface, just like any other protein the cell is making, for examination by patrolling T cells. The viral peptides are quickly recognized by the T cells as foreign. Both CD4 and CD8 cells are activated, a DTH reaction ensues, and the CD8 cells attack the infected cell and destroy it, depriving the viruses of a place to replicate. If they spill out and infect neighboring cells, those cells too will be killed.

The problem is that the immune system has absolutely no way of knowing whether the virus invading a cell is harmful or not. Killer cells make no distinction between cells invaded by pathogenic tubercle bacilli and cells invaded by harmless HBV. They have been selected over evolutionary time to simply destroy *any* cell inhabited by *anything* that is not self, on the plausible assumption that if it is not self, it might kill you. So the immune system is basically blind; it is incapable of making decisions about good and evil, and errs on the side of caution—commendable caution, most of the time.

In HBV infections, most infected cells meet precisely the fate just described. In the acute form of the disease, the response by antibodies and T cells is vigorous, and the infection is usually completely cleared. The resultant immune damage to the patient's liver can be serious, but it is repairable and only rarely fatal.

But in about 5% of cases, the disease may not be resolved at the acute stage and progresses on into the chronic form of hepatitis. This is where the greatest damage is done. The viral DNA continues to direct production of low levels of viral proteins, which make their way to the surface of infected liver cells. And CD8 cells just keep on killing infected liver cells. We see exactly the same game of hide and seek, search and destroy as we saw in the case of chronic TB. The CD8 cells also release inflammation-promoting cytokines like IFN-γ, which just makes things worse. Eventually, all of this can lead to a state called **cirrhosis**, which is a general term referring to massive liver cell destruction. It is a bit like the "liquefaction and cavitation" reaction seen in tuberculosis and is caused by the same thing: relentless destruction by T cells.

The damage in viral hepatitis is similar in outcome to that seen in alcohol- or drug-induced cirrhosis of the liver. Because the liver (uniquely among tissues in the body) has a certain capacity for self-regeneration, the damaged liver constantly tries to replace damaged cells with new ones. But these, too, become infected as HBV spreads slowly throughout the liver, creating an ongoing cycle of destruction and renewal. Unfortunately, over time the renewed liver tissues become more and more abnormal, failing to carry out routine functions such as metabolism of food, removal of toxins, and production of blood coagulation products and bile. In some cases destruction simply outpaces renewal, leading rapidly to liver failure and death. In other cases the constantly replicating liver cells become cancerous and start to grow rapidly and without control. In a high percentage of advanced cases, particularly in third-world countries where the necessary intensive care is unavailable or inadequate, hepatitis viruses are a major cause of liver cancer. The result is usually death.

Bacterial Sepsis—Too Much or Not Enough?

In both tuberculosis and infection by the hepatitis B virus, serious damage to the host occurs when the immune system fails to clear the initial infection quickly and cleanly, allowing it to evolve into a chronic interplay between the immune system and the microbe. Death, when it comes in these cases, is due to the complete failure of a particular organ.

Bacterial sepsis occurs when the body fails to clear an initial bacterial infection that begins locally and then spreads throughout the body. While sepsis often and most lethally starts in the lungs, the sequence of harmful events there is quite different from tuberculosis, because the bacteria involved in sepsis do not invade host cells, but remain free in body fluids. Regardless of where it starts, the fact that the infection is confined to body fluids is what allows it to spread if not quickly suppressed, dragging the immune system's accompanying inflammatory response with it, until literally the entire body may be inflamed.

Bacterial sepsis is a serious problem. In the United States, sepsis afflicts nearly 800,000 people each year, and at least a third of them die of it, making bacterial sepsis one of the top 10 causes of death. It is *the* chief cause of death in most intensive care units, where sepsis mortality rates are often 50% or more.

The major defense against bacteria in sepsis involves elements of the innate immune system, through inflammation, and antibodies plus complement. Killer cells are not involved, since there is no invasion of host cells. The ultimate cause of death, which is often reported as septic shock or total organ failure, is in most cases not entirely clear. But in humans, the ultimate target of damage in bacterial sepsis may be the immune system itself.

During bacterial sepsis, levels of proinflammatory and inflammatory cytokines reach astronomical levels in the blood. Driven by ever-increasing numbers of bacteria, every cell of the innate and adaptive immune systems is fully activated over time, releasing huge amounts of cytokines such as tumor necrosis factor α(TNF-α) and interleukin-1 (IL-1). Toxic degradative products also spill out from hyperactive macrophages and neutrophils. Body fluids contain very high levels of activated complement fragments.

Any or all of these features could be damaging to normal body tissues. Indeed in mice, high levels of TNF-α can be directly toxic to normal cells, inducing them to commit suicide. Tissue damage by digestive enzymes and vascular problems triggered by cytokines are quite substantial. These factors may well explain septic mortality in rodents, since interference with these inflammatory mechanisms can greatly alleviate the damage done by sepsis. So the model of death in mice, due to collateral damage from a "cytokine storm," was assumed to be operative in humans as well. Bacterial sepsis was thought to be yet another example—like TB and viral hepatitis—of the body's immune mechanisms going way beyond what would be required to clear the infection and taking out the host in the process.

But interestingly, in humans, treatments based on tamping down the raging inflammatory response accompanying at least the early stages of bacterial sepsis have at best a minimal effect, and in many cases make the situation worse. This has long puzzled physicians who must deal with human sepsis. Recent studies suggest that in humans, while initial tissue damage caused by rampant inflammation may be harmful, it is by no means lethal. Patients who experience rampant inflammation as the immune system tries to clear the infection, but then recover through antibiotic treatment, suffer minimal or no organ damage. At autopsy, persons actually dying of bacterial sepsis also show relatively minor damage to organs and tissues. Only about 3% to 4% of patients with even severe sepsis progress to septic shock, caused by cytokine-driven enlargement of blood vessels, with concomitant lowering of blood pressure. However 30% to 50% of these will die.

There is one important exception to the general lack of tissue damage. In humans, after an initial burst of cytokines and the inflammation this triggers, there is a rapid reduction in the amounts of several key white blood cells—namely CD4 T cells, B cells, and dendritic cells. The cells that are not physically removed are barely active in carrying out their immune functions. These are precisely the elements most critical in the body's response to bacterial infection. The numbers of CD8 T cells, macrophages, and neutrophils are not reduced, although they too appear to be hypoactive.

Thus, what we see in humans as sepsis develops is an initial burst of inflammatory activity followed in most cases by a crippling disappearance of immune cells that are critical to mounting an inflammatory response, the body's principal defenses against bacterial infection. Contrary to earlier thinking, it has not been possible to attribute the lethal effects of sepsis to a collateral attack by the immune system on self tissues. It seems increasingly likely that the damage seen in bacterial sepsis may be caused not by an attack of the immune system against the rest of the body's tissues and organs, but against itself. In the face of massive bacterial infection, our body's major immune defense—inflammation—appears to be self-extinguishing!

This is a particularly pernicious form of immunopathology. The mechanisms involved are unclear but are presumed to be based on cytokines. It is possible that the cells have been driven so hard in response to the cytokines they produce that they simply become exhausted and collapse—become **anergic**, as immunologists like to say.

In each of the three cases we have examined—tuberculosis, hepatitis B infection, and bacterial sepsis—we find ourselves asking: why does our own immune system do this to us? How could we spend millions of years of evolutionary time and energy and come up with a system that does us so much harm?

Part of the dilemma for the immune system may well have its origin in our success as a species in other areas. Barely a hundred years ago, the immune system had to work time-and-a-half just to keep us alive long enough to reproduce. Today, most of the diseases the immune system evolved to protect us against can be controlled to a large extent by other means, such as hygiene, public health measures, or antibiotics. Damaging overreactions to harmless microbes may today seem like serious medical problems. But over the past half-dozen centuries or so, as humans gathered ever closer in cities and towns, greatly facilitating the spread of infectious diseases in many cases picked up from their domesticated animals, annoyances caused by an overactive immune system may have been scarcely discernible. Given what the immune system has had to overcome to get us to this stage in our evolutionary history, the problems that we now call immunopathologies can hardly be used to label the immune system a failure.

Another dilemma for the immune system lies in the way it was designed. In applying its force, the immune system is, as we have seen, essentially blind. With a few useful exceptions, it has no way of knowing whether a microbe that has invaded the body, and possibly taken up residence inside a cell, is potentially pathogenic or completely harmless. It simply knows the microbe does not belong there and will relentlessly, blindly pursue it until either it is cleared from the system or until, in extreme cases, the immune response finally destroys the host.

One result of the overall success of our immune systems (coupled with other factors like adequate nutrition, isolation from predators, and a generally safe natural environment) is a greatly extended average life span. In the past 100 years, although the maximum human life span still appears to be fixed at 120 years or so, the *average* life span has nearly doubled, due largely to reduction in mortality from childhood infectious diseases. Most animals in the wild live only a short time beyond the peak breeding years for their species.

It may not be very flattering to our egos, but nature does not really have a role for any of us beyond passing on our genes—beyond simple reproduction. We in fact become a potential problem for the next generation of breeders and their offspring by consuming valuable resources needed by younger members of the species for reproduction. The immune system, like other life-support systems, is designed to protect us up through our active breeding and child-rearing season in life. It hasn't the foggiest idea what to do with us beyond that. By and large, many of the problems caused by the immune system as we grow older (such as hypersensitivities, chapter 10; and autoimmune diseases, chapter 12) would be unknown, or at best very minor inconveniences, if we simply cashed out when nature intended—which is hardly an attractive solution to the problem!

7

Vaccines

HOW THEY WORK, WHY THEY SOMETIMES DON'T,
AND WHAT WE CAN DO ABOUT IT

The practice of vaccination was introduced by Edward Jenner in the late 1700s, when he exposed people to cowpox to protect them from smallpox. He did this based on empirical observation. He had noticed that milkmaids would sometimes develop mild fever and smallpox-like blisters on their hands (the condition known as cowpox), but would then be highly resistant to the more deadly smallpox. So he decided to scratch a small amount of pus from a cowpox lesion directly into the skin of a healthy individual, and then later expose that individual to pus taken from a real smallpox blister. Obviously, human subjects protection committees were not yet up and running in Jenner's day! But in the vast majority of cases the individuals he treated were indeed made resistant to smallpox. The practice of vaccination for smallpox using Jenner's original method spread quickly after that.

What Jenner did not—could not at the time—know was that cowpox and smallpox are caused by two closely related viruses.[1] Because the two viruses share many molecular features, exposure to one produces cross-reactive immunity to the other. But not understanding the basis for what he was observing, he didn't know how to apply it to other diseases.

---

1. Jenner's original smallpox vaccine was made from cowpox, but somewhere between Jenner's time and about 1860, probably as a result of experimentation now lost to history, the source for the vaccine changed, and a closely related pox virus called *vaccinia* began to be used. Vaccinia is very close genetically to *V. major*. Its exact origin is unknown, but it may have been isolated from horses.

Advances in our knowledge of microbial biology, some 80 years later, set the stage for a veritable explosion in the widespread application of immunization procedures that would reduce death rates from infectious disease so dramatically in the early twentieth century. When we left this story in chapter 2, Robert Koch had demonstrated that antiserum made against a microbe in a horse (for safety reasons and for economy of scale) could be passively transferred into a human being and overcome microbial infection. But there was one minor and one major problem with this **passive immunization** procedure that limited its use. The minor problem was that while the initial transfer could have a powerful effect, in many cases stopping the infection completely, the transferred antibodies disappear from the blood in a few days and the protection they provide is gone. There is no long-term immunity and no memory with this form of immunization.

The more serious problem was that the antiserum was from a horse and full of horse proteins. Being foreign, these proteins (including the protective antibodies) provoked a powerful immune response in the recipient one was trying to protect. The initial transfer into the recipient worked because it takes the immune system several days to build up a hefty response to the horse proteins. That's enough time for the incoming antibodies to wipe out the infection. But if an additional transfer was needed, the recipient (being now immune to horse serum proteins) mounted an immediate and powerful antibody attack on the antiserum itself.

Not only did this neutralize the protective antibody being administered, but it could also result in huge amounts of antigen–antibody complexes, composed of all the horse serum proteins as antigens and the recipient's antibodies to them, building up in the blood and causing another immunopathology: **immune complex disease**. This happens when the macrophages, which would normally clear these complexes from the blood, are simply overwhelmed by the amounts of complex they have to eat. The excess complexes then settle out in the kidneys, lungs, joints, and elsewhere, where they trigger a host of health problems for the person being treated. We will see a similar problem arising when we look at autoimmune diseases (chapter 12).

So basically, and to this day, passive transfer of immunity is a one-shot deal. It is still used in a number of situations—if you have a particularly massive bacterial infection, it can often be knocked down first with antiserum and then cleaned up with antibiotics. Or you may have been bitten by a poisonous snake, where speed of clearance of the toxin from the body is of utmost importance. It could also be the only remedy in case of a bioterrorist attack with deadly pathogens (chapter 14). But at present, once a particular antiserum has been administered, the same form of that antiserum cannot be used again.

Passive transfer of immunity, by the way, is not just a laboratory or clinical manipulation. It happens in nature. Antimicrobial antibodies are passively transferred to the fetal circulation across the placenta in all mammals. Breast milk (the only kind most mammals get!) likewise contains a wide range of protective antibodies. The first milk that comes from a mother's breast, called the **colostrum**, is absolutely chock full of premade antibodies. These maternal antibodies provide much-needed protection for the newborn until his or her own immune system is up and running and filling in its own memory banks.

But in the larger world, passive transfer of premade antibodies is no substitute for **active immunization**. Active immunization is what happens when we are infected by a microbe and respond to it with the full range of our innate and adaptive immune mechanisms. It was what Jenner was doing in his cowpox/smallpox experiments. The initial response lasts as long as we need to clear the infection. Most importantly, we develop long-term, easily recallable immunological memory to the provoking antigen.

The insight that allowed Jenner's approach to be extended to other diseases came about through a laboratory mistake. The story is somewhat apocryphal, but is thought to have played out in Pasteur's laboratory. Someone studying cholera infection in chickens allegedly left a tube of cholera bacteria sitting out on a benchtop during a heat wave. When that person subsequently tried to use these bacteria to produce cholera in a group of chickens, no disease developed.

The researcher apparently assumed the bacteria had been killed by the heat, so these same chickens were subsequently reinjected with a strong dose of fresh cholera bacteria. Again, there were no signs of disease, which was a bit of a surprise. But *the same dose of this second batch of bacteria caused serious disease in a group of untreated chickens*! Whoever was doing this could have just tossed the whole thing in the can, gone home for dinner, and started over the next day. Luckily, this didn't happen. A blood test quickly showed that chickens receiving the heat-treated bacteria, though not developing disease, had in fact mounted a strong antibody response against the dead bacteria and developed memory B cells. This was *very* effective against the subsequent dose of live bacteria and could block the development of cholera. Pasteur subsequently showed that the same thing could be done with anthrax and rabies.

This was truly the break everyone had been hoping for. It was soon realized that dead or disabled pathogenic microbes, while unable to cause disease, could induce a perfectly good, highly specific immune response. From what you already know about how the immune system works, this makes perfectly good sense. Microbes are composed of exactly the same molecules when they are freshly dead as when they were alive. The immune system has no way of knowing whether they are part of a live or dead organism. It just responds to them as foreign.

The idea of conferring immunity to a deadly pathogen by deliberately exposing someone to that pathogen, even when dead, was not an easy sell at first. Many people fiercely resisted campaigns to carry out immunizations on a public health scale. But we now produce vaccines for immunizations against a wide range of microbes either after they are killed or after their pathogenicity has been greatly **attenuated**. Microbes can be attenuated by various means. They can be grown under conditions that favor mutation, and mutational variants that lose their **pathogenicity** while retaining their **immunogenicity** can be isolated and grown out in large quantities for use in vaccines. For some microbes, it has also been possible to prepare vaccines that consist, not of the entire organism, but of key molecular structures isolated from the microbe in the laboratory. Isolated and inactivated bacterial toxins can also be effective and useful vaccines.

Admittedly there were some tragic miscalculations along the way. Even now, on rare occasions someone will have an adverse reaction to a vaccine for reasons that are not clear. But we routinely administer a large number of these vaccines to infants in their first two years of life, for one simple reason. The odds of a child getting a serious infectious disease as a result of vaccination is literally thousands of times less than the odds of getting the disease from an unimmunized population of his or her peers. This is the double benefit of immunization. Not only are individuals protected from a particular microbe by bolstering their immune systems, but also when populations are immunized, individuals—even those *not* immunized—benefit from the huge reduction in the frequency of that microbe in the vaccinated general population.

The goal of all vaccination is to mimic the immunological consequences of a natural infection by the corresponding microbe. One of the outcomes of a natural infection is the selective expansion of T and B cells that recognize the microbe and the conversion of some of these to long-lived **memory cells.** These expanded memory cells respond more quickly to the microbe the next time it comes into the system. This selective expansion of highly reactive cells gives the vaccinated individual a much greater chance to overcome a subsequent infection *before* infectious disease symptoms or microbial immunopathologies can develop.

Active immunization has now led to the complete eradication of smallpox in the world. Samples of the smallpox virus are currently stored frozen in the United States and Russia (safely, we hope) for future study, should that be necessary. A campaign managed by the World Health Organization (WHO) is attempting to eradicate polio as well. While there have been some setbacks, it seems likely that within a few years this disease, too, will become a thing of the past. Measles has been all but eliminated in industrialized countries, but a vaccine stable enough for delivery in third-world countries has yet to be developed. More than a thousand children a day still die of measles in underdeveloped countries—not for lack of a vaccine, which has existed for decades, but for lack of a *stable* vaccine.

The road to our present state of protection through planned vaccinations was not an easy one. It took over a hundred years after discovery of the typhoid bacterium before we had an effective vaccine against it, and nearly that long to develop a flu vaccine. There are still dangerous pathogens for which no effective vaccine exists. The reasons for this are many and varied. The WHO has recently recognized three infectious diseases for which an effective, readily deliverable vaccine has still not been produced, and which in combination account for a very large proportion of deaths from infectious disease worldwide: malaria, tuberculosis, and HIV/AIDS. Combined, these three diseases kill 6 million people each year. Other diseases desperately needing an effective, long-lasting, stable vaccine include measles, tape worm, and flu.

We will look at malaria and tuberculosis here, from the point of view of vaccine development. We will have a more detailed look at AIDS vaccines in the next chapter. Based on many of the things you have already learned, there are some exciting new approaches to making vaccines that promise to revolutionize this life-saving procedure over the next few years.

MALARIA

Malaria is an infectious disease caused by neither a bacterium nor a virus, but by a protozoan parasite, *Plasmodium falciparum*. Protozoan parasites are also single-cell organisms, but represent a much larger and more complex form of unicellular life. These kinds of parasites are responsible for other infectious diseases, such as

African sleeping sickness (*Trypanosoma sp.*), **encephalitis** (*Toxoplasma gondii*), and **leishmaniasis** (*Leishmania major*).

The WHO estimates that about 500 million people worldwide are infected with *P. falciparum* or its relatives annually, and 2 million die of this disease each year. The vast majority of cases are in sub-Saharan Africa, where one child in five dies of malaria. Death is slow in coming and can be preceded by many years of debilitating disease. The disease can be controlled to some extent by drugs, but these are expensive and must be taken over long periods of time, both of which compromise drug treatment in developing countries. In some areas the parasites also become resistant to many effective drugs, such as chloroquine.

The malaria parasite has a complex life history in its host, infecting at different times both red blood cells and liver cells, and in some cases the nervous system as well. The immune response to the malaria parasite takes different forms at each of these stages. Parasites enter the body through the saliva of a feeding mosquito as it feeds and make their way quickly to the liver, where they hide inside cells while reproducing. CD8 cells will attack infected liver cells, just as if they were infected with a virus or a bacterium. Cells not killed by CD8 T cells eventually burst open, and the parasite leaves to find red blood cells for its next stage of development.

Antibodies now play the major role in defense. They can neutralize the parasites in transit. Also, as the parasites enter the red blood cell, they deposit some of their proteins on the cell surface. This is not due to transport of peptides to the surface with MHC proteins. These parasite proteins are spotted by antibodies, which either initiate complement-mediated killing or tag the parasites for phagocytosis by macrophages. But parasites surviving in the blood stage of infection can be taken up by another feeding mosquito and passed to a new victim.

In most cases, even with repeated infections from childhood onward, neither antibodies nor killer cells are entirely able to clear a natural infection by *P. falciparum*. Moreover, in addition to the damage done by the parasite itself, in the liver and in the blood,

there is a significant component of chronic inflammatory immunopathology contributing to the overall pathology of malaria. This
is particularly true of the damage done to the nervous system. CD8
cells also produce IFN-γ, activating a strong macrophage response,
which in turn boosts the inflammation. But since the infection never
entirely clears, the inflammation and the accompanying damage
just go on and on.

One contributing factor to the inability to clear infections is that,
like the influenza and AIDS viruses, this parasite can rapidly mutate many of the proteins detected by antibodies. Perhaps most importantly, long-term memory does not seem to develop in infected
individuals, so when someone is reinfected, the response is not much
stronger than when that person was infected for the first time.

This is the major challenge for developing a malaria vaccine.
Since we don't understand why memory does not develop during
the course of a natural infection, it is difficult to know exactly how
to overcome this problem with a vaccine. In general, it is felt that
a stronger CD8 response during the liver stage of the infection
would be the best contribution from a vaccine, since CD8 cells are
involved in the earliest stages of the infection, both through IFN-γ
release and as killer cells.

TUBERCULOSIS

We have already seen what tuberculosis does to the body, both in
terms of damage by the bacterium itself and the accompanying
immunopathology. It is a major human health problem that was
thought to be under control through vaccination, but is once again
on the increase throughout the world. Currently, 30% of the world's
population is thought to be infected. Three million people die of it
each year.

Tuberculosis can be managed reasonably well by drugs, but this
only works in industrialized countries where people have ready

access to the drugs and where long, complex treatment programs can be monitored by a physician. The vaccine that contributed to the major reduction in this disease in Europe and the United States was developed over 100 years ago. For reasons that are not clear, this vaccine has worked less well in the third-world countries. It works well in children, but wears off in about 10 years, and does not work particularly well in adults. There is also concern that new, drug- and vaccine-resistant strains of M. *tuberculosis* may be emerging in connection with the worldwide AIDS epidemic.

One of the goals of an improved tuberculosis vaccine will be to induce a stronger, longer-lasting state of immunity, and in particular immune memory. Moreover, it is generally agreed that CD8 T cells play an important role in the early stages of the disease. If more killer cells could be brought into play, although perhaps causing some severe short-term collateral damage, they might well be able to clear the infection.

NEW APPROACHES TO VACCINE DEVELOPMENT

For humans, the most critical role of a vaccine will be to enhance the adaptive immune response to any microbial pathogen. The most critical cell in mounting an adaptive response is the T cell, since these are crucial in both antibody responses and in killer T-cell development. However, recently developed knowledge of how the innate immune system functions to get the adaptive response up and running offers some of the best approaches for new, more effective vaccines.

For the two diseases we have just looked at, malaria and tuberculosis, and for all viral diseases, the CD8 cell response is particularly critical, and most new vaccine strategies are focused on enhancing this arm of adaptive immunity. But as can be seen from Figure 4.1, CD8 T cells for the most part are activated in response to microbial proteins actually manufactured within a cell. If we use killed or disabled viruses, or proteins isolated from viruses (or any other intracellular pathogen), there will be minimal activation of CD8 cells.

Dendritic cells play a key role in the response to every microbial pathogen studied and are by far the most effective in presenting antigen to both CD4 and CD8 T cells. Dendritic cells are thus at the heart of most recent strategies to develop vaccines for viruses and other intracellular parasites. Dendritic cells are able to present high concentrations of MHC-associated foreign peptides at their surface, where they also display a wide range of activating coreceptor ligands and provide cytokines needed for T-cell maturation. So increasing the presentation of antigen through dendritic cells, and ensuring that the dendritic cell itself is optimally activated, is the most promising pathway for boosting CD8 T-cell responses.

There are many ways of delivering antigen to dendritic cells. Dendritic cells can be isolated from individuals and incubated in the laboratory with peptides from a particular microbe that are known to induce a good immune response. The peptides, if presented in a high enough concentration, will displace peptides already associated with class I molecules on the surface of the dendritic cell. This does not require the peptides to be internalized by the dendritic cell. These "pulsed" cells can then be reinfused back into their owner and will find their way to lymph nodes, where they present the pulsed peptide to T cells.

The most interesting and potentially most powerful way to get peptides into dendritic cells is not to feed them peptides at all. Implant in them instead the *gene encoding the peptide* you are interested in. This is what is happening in the exciting new field of **DNA vaccination**. A piece of DNA containing the gene specifying the synthesis of a given peptide, which we know from other studies is recognized as foreign by CD8 cells, is simply introduced into the body at a site known to be rich in dendritic cells. These are in fact the sites normally used for vaccination as we already know it— the skin and muscle.

The DNA will be taken up by a dendritic cell and in a significant number of cases will make its way into the nucleus and be read just as any other gene in the cell. The cell will then produce the corresponding protein. It's as if the microbe itself were active inside the cell, directing the synthesis of its own proteins for purposes of reproduction. But the whole microbe is not there—just one of its genes we have selected for purposes of vaccination. The corresponding peptide will appear at the surface of the dendritic cell, bound to its transporter class I molecule. And because it is foreign, it will activate any CD8 cell that recognizes it.

This process of **transfection** of dendritic cells with genes encoding potent microbial antigenic peptides has proved very effective in laboratory studies with animals, and DNA vaccination is now being used in clinical trials for humans. However, with the possible exception of AIDS vaccines, it may be several years before DNA vaccines for human disease become widely available even in the industrialized countries that are producing them. But DNA vaccines have recently been developed that prevent infection of monkeys by Ebola and Marburg viruses. Human trials should follow within the next few years.

Another exciting development that is increasing the efficacy of vaccines, for both peptide pulsing and DNA vaccination, takes advantage of our knowledge of the cytokines needed by various immune cells as they undergo activation. The dendritic cell used can be given a gene encoding a microbial protein *plus* an extra gene for one of the cytokines needed by the responding T cells. These cytokines will be produced and secreted into the immediate vicinity of the dendritic cell and provide a highly focused exposure of the T cell to these much-needed growth factors. In the case of DNA vaccination, a cytokine gene can be attached directly to the peptide gene that is being used for immunization. We will see a variant of this approach when we discuss vaccines for cancer in chapter 11.

These are but a few of the ways that vaccines are being made more effective based on recently acquired information about how the innate and adaptive immune systems interact with one another.

DNA vaccines will almost certainly be the way of the future for many human vaccines. Not only can they be tailored toward particular antigenic microbial molecules, but they may also solve some of the logistical problems that have plagued vaccination programs in developing countries. Most current vaccines require constant refrigeration if they are to last more than a week or two. Given the problems of shipping these vaccines to distant spots on the globe and the lack of infrastructure in many countries to provide refrigeration outside of large cities, it has been impossible to bring in vaccines that could save millions of lives each year. A great advantage of DNA vaccines is that, properly prepared, they are for all intents and purposes indefinitely stable. They can be carried around in an aid worker's shirt pocket or purse for weeks, if necessary, to reach remote areas.

For even DNA vaccines to be fully effective, however, it will be necessary to move beyond the standard needlestick approach to vaccine delivery. The repeated use without sterilization of needles in vaccinations throughout third-world countries has itself become a major factor in the spread of infectious disease. For example, the WHO estimates that at least 20 million new cases of hepatitis B are created each year by passage of the virus from infected to noninfected people through dirty needles in vaccinations unrelated to hepatitis B. (There is an effective vaccine for hepatitis B used in industrialized countries, but for reasons of cost as well as logistical problems, this vaccine has had a minimal impact in developing countries.) Noninvasive approaches to vaccination, such as air guns or skin patches, are desperately needed. But once the vaccines are ready to go these are, in the end, relatively minor technical problems.

IMMUNOLOGY MEETS ECONOMICS

The unraveling of all the mysteries underlying how the immune system works and how deliberate immunization—vaccination— can produce stable states of immune resistance to defined pathogens has been carried out largely in university research laboratories, in studies funded by government agencies or private foundations. But such laboratories are not equipped, either by temperament or logistical infrastructure, to convert this knowledge into large-scale efforts to develop vaccines and distribute them to entire populations. This has been the province of the pharmaceutical industry. The pharmaceutical industry has the scientific and engineering know-how, the physical facilities, and the financial and marketing skills to apply the information gathered by tens of thousands of individual research scientists to a safe, practical medical product with the potential to save millions of lives.

This partnership between basic science and private industry has worked very well in developed countries, resulting in vaccines that have led to the eradication or near-eradication of infectious diseases that barely a hundred years ago killed tens of millions of people each year. But this partnership has one limitation. Pharmaceutical companies can only survive and function if they make a profit. Bringing a vaccine to market, especially in the heavily regulated drug industry that has developed over the past half-century, requires an enormous investment of time and money. Drug companies simply cannot absorb these costs without some hope of return. For many vaccines, such as flu, polio, measles, or any of several dozens of other infectious diseases impacting industrialized societies, there is ample opportunity for reward for companies willing to risk the investment.

But for infectious diseases that do not affect significant numbers of people in developed societies, few companies are willing to take that risk. University researchers may work out pathways for the correction of such diseases, but if their work uncovers an effective drug, the investment required to bring such a drug through the necessary clinical trials and Food and Drug Administration reviews could never be recovered in an open market system. Alternatively—and this is the case with vaccines—there may be a huge potential patient base, but not a patient base that can afford the free-market costs of the treatment.

TABLE 7.1
The Global Alliance for Vaccines and Immunization (GAVI)

DONOR NATIONS	INDUSTRIAL PARTNERS	INTERNATIONAL GROUPS	PRIVATE FOUNDATIONS
Canada	Chiron	UNICEF	Gates Foundation
Denmark	Glaxo SmithKline	WHO	
France	Wyeth	World Bank	
Ireland	Merck	UN NGOs	
Netherlands			
Norway			
Sweden			
European			
Luxembourg			
United Kingdom			
United States			

The world community has been struggling with this dilemma for many decades, and it now seems likely that this problem can be solved through the combined and coordinated efforts of scientists, governments, and the private sector, including both pharmaceutical companies and private philanthropy. One of the most exciting of these new combines is something called **GAVI**—the **Global Alliance for Vaccines and Immunization** (www.vaccine alliance.org; Table 7.1). Created in 2002, this privately assembled international body coordinates the efforts of international organizations such as the WHO, the United Nations and its various nongovernmental organizations (NGOs), the World Bank, national governments, major drug companies, and private foundations such as the Bill and Melinda Gates Foundation. This unprecedented focusing of private, governmental, and international body attention and money on the problems remaining in vaccine development are certain to yield major dividends in the coming decade.

8

When the Wall Comes Tumbling Down
HIV/AIDS

AIDS was first reported in 1981 by a group of young physicians at UCLA, who noted five cases over a short time period of *Pneumocystis carinii* pneumonia in five previously healthy men. *P. carinii* is a fungus and is what is known as an **opportunistic pathogen**. These are disease-causing microbes that infect someone and provoke a good immune response. But the immune response does not drive the pathogen completely out of the host. Rather, the pathogen lives in some sort of balance with the host's immune system, kept at subclinical levels most of the time but occasionally rearing its head, only to be slapped down again. The herpes virus that causes cold sores is a common example. But these pathogens can grow out to dangerous proportions in persons whose immune systems have somehow been compromised. They are a common complication of the immunosuppression given to transplant patients, for example (chapter 13).

The fact that this form of pneumonia appeared in a cluster of five men who had no apparent reason to be immunosuppressed was so striking that the physicians immediately reported it to the Centers for Disease Control (CDC) in Atlanta. Barely one month later it was reported that 26 male homosexual patients seen during a short time period in another state had *P. carinii* pneumonia and/or Kaposi's sarcoma, a rare form of skin cancer. The UCLA cluster, upon examination, also turned out to be gay. In a matter of months, new cases belonging to this syndrome numbered in the hundreds, and then the thousands, and it had acquired a name: acquired immune deficiency syndrome (AIDS). Thus began a journey into the unknown, a journey whose end is still beyond our vision.

When HIV, the human immunodeficiency virus causing AIDS, was identified in 1983, everyone breathed a deep sigh of relief. It was a virus of a generally recognized type—a retrovirus. Virologists already knew a fair amount about retroviruses. Surely a means for stopping or at least controlling this disease must be just around the corner.

But only months later came a heart-stopping announcement: 18 of the first 19 HIV isolates taken from AIDS patients were immunologically different. What that meant was that the coat proteins in which HIV wraps itself, the parts of the virus detected by antibodies, must be mutating at an extraordinarily high rate. In practical terms it meant that it would be very difficult to prepare an antibody vaccine against HIV. Any form of the virus killed and used as a vaccine one month would probably induce perfectly good antibodies against that strain of HIV, but the resulting defense would be useless against forms of the virus floating around the next month.

This has proved all too true. The same is true of colds caused by the flu virus, which also changes its coat proteins, although not at so fast a rate. That is why there is no vaccine that protects us against all forms of the flu, and why we never build up immunological memory from one cold that can protect us against the next wave of flu virus a year later. No one has ever tried very hard to solve this problem, because usually our own immune systems manage to get on top of each influenza infection and rid our bodies of it. But this doesn't happen in AIDS.

AIDS AS A MEDICAL PROBLEM

How big a medical problem is AIDS? As 2005 drew to a close, the World Health Organization estimated that more than 40 million people worldwide were HIV infected or had full-blown AIDS. At least 25 million have already died. Fourteen thousand new AIDS cases are added each day. Ninety-five percent of HIV-infected individuals worldwide are heterosexual, and 50% are women. In developing countries, only about one in five infected persons receives treatment. All but a handful of these people will die of their disease; two-thirds of all the people ever infected with HIV have already died. AIDS is now the number one cause of death by infectious disease and the fourth leading disease cause of death throughout the world.

The overall shape of the HIV/AIDS pandemic in the United States is shown in Figure 8.1. Currently more than 1 million people in the United States are HIV infected or have frank AIDS. Only about a third of those infected are aware they are infected. Three-quarters of affected individuals are male, and half of those are black. Male homosexuals now account for less than half of existing cases. Current estimates for new cases range between 40,000 and 60,000 annually.

How do we make sense of numbers like these? What sort of perspective can we put them into? On the one hand, AIDS hasn't

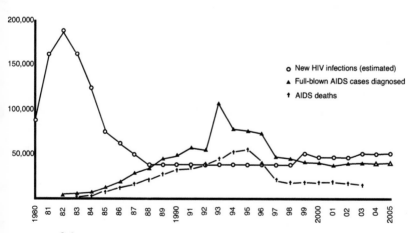

FIGURE 8.1
The HIV/AIDS pandemic in the United States.

yet come close to something like smallpox or hepatitis B as a killer of human beings on a historical scale. It is coming close to, and may have already surpassed, the influenza epidemic that swept 30 million human beings from the face of the earth just after World War I. What is so frightening about AIDS is the speed with which it is spreading, the incredible rate of increase in the number of cases diagnosed each year, with absolutely no cure in sight.

HIV, which is passed almost exclusively by exchange of body fluids, infects, and causes a gradual loss of, CD4 helper T cells in infected individuals. When the level of circulating CD4 T cells falls below roughly half the normal value, HIV-infected individuals begin experiencing the symptoms of **frank AIDS**, including infection by external pathogens and by internal opportunistic pathogens.

There are currently 18 drugs approved by the Food and Drug Administration (FDA) for treating AIDS (see Table 8.1), used alone or in various combinations. They fall into four categories: nucleoside analogs, reverse transcriptase inhibitors, protease inhibitors, and a fairly new category, fusion inhibitors.[1] We will see how these drugs work in a moment. These drugs can be enormously expensive, barely within the capacity of industrialized medical care systems to fund, and out of the question, at least at market prices, for developing or third-world countries.

Current standard therapy in the United States and other industrialized countries is something called **HAART** (highly active antiretroviral therapy). This consists of a combination of at least two reverse transcriptase (including nucleoside analogs) inhibitors, plus a protease inhibitor. Slight variations of this formula are also possible. The idea is to force the virus to mutate to multiple drugs at the same time in order to reproduce, which is much

---

1. Nucleoside analogs are in fact also reverse transcriptase inhibitors, so there are only four distinct categories. But for historical reasons, the two drug types are usually referred to separately.

TABLE 8.1
Current Drug Therapy for AIDS

DRUG	ACTION	APPROVED	COST/YEAR
Zidovudine (AZT)	Nucleoside analog	March 1987	$4,300
Didanosine (ddI)	"	October 1991	3,700
Zalcitabine (ddC)	"	June 1994	3,000
Stavudine (d4T)	"	June 1994	4,100
Lamivudine (3TC)	"	November 1995	3,800
Abacavir	"	December 1998	4,900
Tenofovir	"	October 2001	5,200
Emtricitabine	"	July 2003	3,600
Ritonavir	Protease inhibitor	March 1996	9,000
Indinavir	"	March 1996	6,300
Saquinavir	"	December 1995	5,200
Nelfinavir	"	March 1997	8,600
Amprenavir	"	April 1999	4,700
Atazanavir	"	June 2003	10,000
Nevirapine	RT inhibitor	June 1996	4,400
Delavirdine	"	April 1997	3,800
Efavirenz	"	September 1998	5,300
Enfuvirtide	Fusion inhibitor	March 2003	20,000

Nucleoside analogs are incorporated into HIV DNA and stop DNA synthesis. *RT inhibitors* block HIV reverse transcriptase (RT). *Protease inhibitors* stop the processing of HIV proteins needed to generate a new virus. *Fusion inhibitors* block entry of HIV into cells. Current therapy is based on combinations of these types of drugs and is called HAART (highly active antiretroviral therapy). Properly administered, HAART can reduce HIV levels in the blood to essentially undetectable levels. The "HAART attack," however, does not cure the HIV infection.

more difficult for individual viruses to do. HAART combinations have greatly reduced the viral load in most patients, but the expense is compounded, as are the side effects. Not everyone can tolerate HAART for long periods of time.

Drug regimens like HAART have made a significant impact on survival with AIDS. In the early 1980s, a person might be expected to live two to three years after onset of frank AIDS. Today, it is not unreasonable to expect to live 8 to 10 years, if all drug regimens are strictly adhered to. That is not easy, of course. Not only is it expensive, but the potent metabolic side effects and the constant attention to complicated times and dosing of medications lead many people to fall off the regimen. The consequences are evident in the accelerated decline of the individual. Unfortunately, for every drug that has ever been developed, the virus eventually mutates to a form resistant to that drug. The heavy toll caused by side effects and the knowledge that there may never be a true cure for AIDS in their lifetime leads many to give up.

AIDS AS A PROBLEM IN VIROLOGY

HIV may now be the most intensely studied virus ever. HIV is an RNA retrovirus, which means that its genetic blueprint is written in the RNA code rather than in the DNA code used by all animal (including human) cells. The entire virus consists simply of this piece of RNA wrapped in a small number of coat proteins plus a few lipids.

Like all viruses, HIV must infect a living cell in order to reproduce. The first step in the infectious process is the binding of HIV to the surface of a cell. HIV doesn't just "stick" to any cell; it gains entry to a cell through specific molecular interactions, like an antibody binding to an antigen. One of the prominent proteins making up the coat of HIV is a **glycoprotein** (a protein that contains sugar molecules in its structure) called **gp 120**. HIV uses gp 120 (the 120 refers to its size in atomic mass units) to bind to the cell it is going to infect. The gp 120 protein specifically recognizes and binds to the CD4 molecule found mostly on CD4 T cells, but to a lesser extent on macrophages and possibly certain brain cells. It is this predilection of HIV to bind CD4 molecules that ultimately makes this virus so deadly (Figure 8.2).

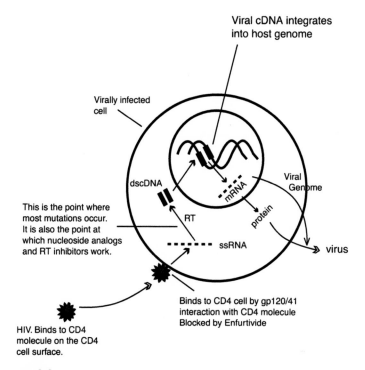

FIGURE 8.2
Infection cycle for HIV. Drugs like zidovudine (nucleoside analogs) get incorporated into DNA, rendering it useless. These analogs are not incorporated efficiently into human DNA, but do get incorporated at low efficiency, causing serious side effects. Drugs like nevirapine block the reverse transcriptase (RT) inhibitor enzyme, but have no side effects.

But simple binding of HIV to CD4 on the cell surface is not enough for the virus to gain entry to the cell. When gp 120 attaches to CD4, the gp 120 molecule twists slightly, revealing a second binding molecule, gp 41. The gp 41 molecule must bind with a second **coreceptor** on the CD4 cell surface, called **fusin**.

Only with both binding events in place—gp 120 to the primary receptor, CD4, and gp 41 to the coreceptor, fusin—can HIV enter the cell (Figure 8.3).[2]

Although this seems a little complicated, it holds the key to a puzzle that had long intrigued AIDS researchers. It was known by the late 1980s that a very small number of individuals, mostly of northern European descent, can be infected by HIV but never develop AIDS. HIV is in their bloodstream, but never gets into their CD4 cells. It turns out that these individuals have mutations in the gene for the fusin coreceptor on their CD4 cells, resulting in a deformed or absent coreceptor. This prevents HIV from entering the cell whose infection would lead ultimately to destruction and frank AIDS.

If HIV ever mutates from a virus that can be passed in aerosol form like the flu virus, so that a sneeze in an elevator can infect a dozen people at a time, individuals with the fusin mutation could be the only human beings to survive. It takes at least four to five years for someone with HIV to develop the symptoms of AIDS. In the meantime, that person can sneeze in a lot of elevators, or at parties, or on airplanes. The rate of spread would be exponential, and it is hard to see how it could be stopped.

Each HIV virion that infects a CD4 cell brings with it a preformed molecule of an enzyme called **reverse transcriptase (RT)**, which it uses to convert its RNA into DNA, the genetic language of the host cell. This is the point where HIV genes mutate like crazy, and also the point where nucleoside analogs and RT inhibitors work (Figure 8.2). Mutations arise here because the RT enzyme is prone to making mistakes in copying HIV RNA into DNA. Moreover, the copy errors are not edited.

---

2. This critical fusion step is blocked by the drug enfuvirtide (Table 8.1). This drug has shown every promising results but requires twice-daily injections, making it very expensive and a regimen difficult for some patients to adhere to. Drug companies are working feverishly to develop forms of the drug that last longer in the blood.

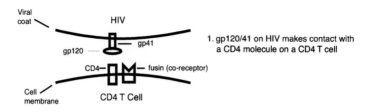

Viral coat

HIV

gp120 — gp41

CD4— fusin (co-receptor)

Cell membrane

CD4 T Cell

1. gp120/41 on HIV makes contact with a CD4 molecule on a CD4 T cell

This is the point where the new drug enfuvirtide works. It blocks gp41-fusin contact

gp120 — gp41

2. As a result of step 1, gp41 partially unwinds into a form that can make contact with the co-receptor *fusin*

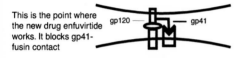

3. As a result of step 2, the membranes of HIV and the CD4 T cell fuse, opening a channel across which the contents of the HIV virion can pass.

FIGURE 8.3
Entry of HIV into a CD4 T cell is mediated by two distinct molecular interactions.

Most of these mutations are likely to be deleterious to the virus, but that doesn't matter. The virus reproduces so rapidly inside a living cell that errors are affordable, as long as a few functional viruses are made in the process. The advantage to the virus is that occasionally one of these mutations produces a new strain of virus that is more effective than the virus that originally infected the cell, perhaps by being more resistant to drug treatment. Mutations in the coat proteins of the virus are particularly important in helping the virus escape destruction by the immune system.

The structure of an HIV virion (the completely assembled virus, ready for another round of infection) is shown in Figure 8.4. Also shown is the tiny HIV genome, which contains only nine different genes. Six of these (vif, vpr, rev, tat, vpa, and nef) are small genes coding for elements that regulate HIV reproduction inside the host cell. Three of the nine genes are rather large genes coding for complex proteins. The **gag** protein contains three smaller proteins used to build the basic structure of the virion. The **env** protein contains the key molecules gp 41 and gp 120, involved in fusion. Finally, the **pol** protein contains three molecules that must be passed forward as part of each newly completed virion: one molecule each of protease, reverse transcriptase, and nuclease.

When a newly completed virion is released and infects the next cell, the preformed RT will be used to convert the virion's RNA genome into DNA, so that it can be inserted into the host DNA. The nuclease molecule cuts open the host DNA, inserts the virus DNA, and closes the DNA cuts. The protease molecule, as we just saw, is used to cut the next round of newly minted complex proteins.

The HIV DNA copied from the infecting virus RNA by RT makes its way into the nucleus of its new host, where it inserts into one of the host chromosomes. From that point on the host cell regards the HIV DNA as part of its own DNA and will follow whatever instructions are encoded in it. Retroviruses are unique in that respect. Most viruses just let their DNA float around loose in the nucleus, where it is also read by the host cell but is eventually degraded. Retroviral DNA is forever. The human genome contains large amounts of old retroviral DNA that has made its way into human DNA and been carried forward from generation to generation. Such "junk" DNA, which includes many other forms of "fossil" and generally unused DNA, comprises over 95% of the entire human genome.

When a new virion is completely assembled, it makes its way to the surface membrane of the cell, using the cell's own transport machinery. At the surface it fuses with the cell's membrane and gently pushes out, making a small "bud." The virion continues to

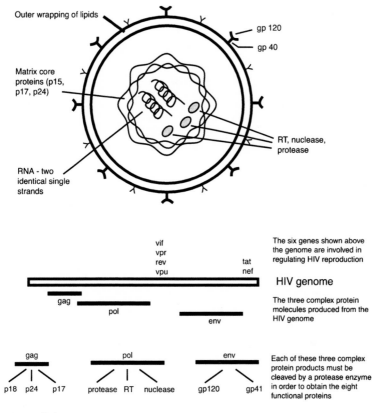

FIGURE 8.4
The structure of an HIV virion and its genome.

push out, picking up some of the lipids in the cell's membrane to complete its own coat. Finally the virion pushes through and floats away in the lymph surrounding all cells. The CD4 cell membrane quickly fills in the tiny hole. This budding process itself causes no damage to the host cell. A new generation of drugs under investigation at several pharmaceutical companies aims to block this budding process, although this may just cause the cell ultimately to burst and release the virions anyway.

THE IMMUNOLOGY OF AIDS

AIDS is not caused by an a priori inability of the human immune system to respond to HIV infection. In fact, there is every sign that the immune system responds as vigorously to HIV as it would to any other virus it encounters. But of the tens of millions of humans infected so far with HIV, there is not a single documented case (aside from those with mutated fusin co-receptors) of someone clearing the infection from his or her body.

We don't have detailed data on the very early stages of HIV infection in humans, because we never see these stages. To do so, we would have to purposely infect a group of individuals and carry out a series of daily tests and measurements. That clearly isn't going to happen. But from information provided by those who eventually turn out to be HIV positive, we know that at the beginning of the infection the affected individual is only mildly symptomatic—a slight fever, a sore throat, or just not feeling "100%," symptoms that accompany almost any viral infection and that we have learned to generally ignore because they almost always resolve. And initially, an HIV infection *appears* to resolve, thanks to the immune system.

In spite of not being able to study HIV directly in the earliest stages of infection, here is what we think is going on, based on studying HIV-like infections in other primates. Once the virus enters the body, it tends to concentrate in those locations where the cells it can infect—CD4 T cells, macrophages, and dendritic cells—reside. This is mostly the various lymphoid tissues and organs, but since these cells are also scattered individually throughout the body (particularly in the gut), so is HIV.

The virus undergoes a brief period of explosive replication while the immune system gears up to fight. Interferons surely must enter the fray early and doubtless slow the virus down a bit. The vast majority of macrophages are not yet infected and will be gobbling up virions as they sprint from cell to cell. After a few days, antibodies are very likely being made, although for a variety of reasons they may not be detected in the serum for up to several weeks. This event, called **seroconversion**, was for some years the first reliable marker that the infection actually existed. There is also considerable free virus in the blood at this time. CD8 killer cells are present in the early stages of the infection, and those who show an early and strong killer cell response tend to have the longest time period before conversion to frank AIDS. It is this all-out attack by the immune system that seems to resolve the initial HIV infection, or at least its symptoms.

The period between seroconversion and the onset of frank AIDS, referred to as the **latent period**, is generally asymptomatic and can be anywhere from 6 to 12 years in adults, less in children. During this period the levels of virus in the blood usually decrease but continue to increase dramatically in lymphoid tissues. The proportion of CD4 T cells, macrophages, and dendritic cells infected with HIV continues to increase, until at some point the balance is tipped in favor of the virus and the immune system begins to collapse, signaling the onset of frank AIDS.

During the latent period, there is a long, drawn-out battle between the immune system and the virus. Antibodies are constantly made and are a major factor in forcing the virus to diversify through mutation in order to escape annihilation. CD8 killer cells are also attacking cells in the body harboring HIV, which in one sense adds to the problem—since HIV itself is killing those same cells—but as in other viral infections, a vigorous CD8 cell attack, depriving the virus of a place to replicate, can keep the infection under control.

The most reliable predictor for the progression of HIV infection toward frank AIDS is the level of viable CD4 T cells remaining in the blood. Most HIV-infected individuals with CD4 counts above 500 (500 cells per cubic millimeter of blood), are usually still asymptomatic. Between a count of 500 and 250, oral candidiasis (a fungal infection of the mouth) and tuberculosis are the most common opportunistic pathogens to rear their heads; at 200 to 150, it is Kaposi's sarcoma and lymphoma that are seen most frequently;

and below 150, deadly opportunistic pathogens such as *P. carinii* and cytomegalovirus make their appearance.

The consequences of CD4 T-cell depletion are complex. CD4 T cells affect virtually every phase of immune responsiveness—and not just antigen-specific components such as other T-cell subsets and B cells. CD4 T cells, through the cytokines they produce, affect almost every component of the innate immune system as well. CD4 T cytokines are involved in communication between the immune system and the brain. In one's wildest imagination, one could not possibly pick a worse cell (from the host's point of view) to serve as the target for an infectious virus.

Strangely enough, to this day no one knows exactly how HIV kills CD4 T cells. We know that within 30 minutes of infection of a cell, HIV has shut down hundreds of host cell genes and activated some of the genes for apoptosis (cell suicide). HIV also causes infected cells to clump with each other and with uninfected cells. The contribution of CD8 killer cells specific for HIV peptides to CD4 cell and dendritic cell loss has been difficult to estimate, but is probably a factor. Any one of these or other mechanisms yet undiscovered could account for the rapid and ultimately fatal loss of CD4 cells. But the fact is that we just do not know.

GENE THERAPY APPROACHES
FOR TREATING HIV/AIDS

After more than 25 years, there is still no cure in sight for AIDS. Over the past 10 years or so, researchers have begun to explore various ways of using gene therapy or other manipulations of DNA (which come under the general rubric of **molecular medicine**) as a means of bringing this infectious disease under control.

Gene therapy as originally envisaged was intended to repair disease-causing genetic defects by introducing good copies of defective genes into cells in which these genes were expressed. The good gene would take over for the defective copy, restoring the gene function in that cell and reversing the cellular malfunction underlying disease. As we will see in the next chapter, this has worked reasonably well for certain genetic diseases of the blood, particularly those involving genetic defects in T cells.

In the case of AIDS, the aim of gene therapy is different. Rather than rescuing a defective cell, we want either to prevent HIV from invading a cell in the first place or, barring that, to prevent HIV from replicating inside a cell it has infected. And if both those attempts fail, we want to be able to destroy an HIV-infected cell before replication can take place.

In the paragraphs that follow, we examine a few of the more interesting possibilities currently being pursued to build a defense system against HIV into human T cells.

The "Entry Denied" Option

Looking back at Figure 8.3, we see that in order for HIV to gain entry to a CD4 cell, the gp 41 molecule on its surface must interact with fusin on the CD4 cell. People who have no fusin on the surface of their CD4 cells are completely resistant to HIV infection. As we said in an earlier context, whenever a microbe has a structure it absolutely cannot alter, that structure becomes a prime target for an immune defense mechanism. Of course, it might take a million years or so for humans to create such a defense. In the meantime, such a structure also becomes a target for therapeutic intervention. There is already a drug on the market (enfuvirtide) that blocks the interaction of gp 41 with fusin. The drug appears to be effective, but it is very expensive, and it would have to be taken essentially every day for life.

A genetic approach to this involves something called **antisense**. The details of how antisense works are beyond the scope of this book, but the idea is simple. In order to place fusin on the surface of a cell, the nucleus sends a message, copied from the fusin gene, out to the cell to direct production of the fusin protein. An antisense gene for fusin can be constructed and inserted into a T cell, for example, and it will make an antisense fusin message to send out to the cell.

But before these two messages—sense and antisense—even get out of the nucleus, they collide and neutralize each other. It's the same principle as matter and antimatter—not as spectacular, per-haps, but potentially as effective. We feel confident this wouldn't cause a problem, because those lucky few who lack this molecule entirely in their body do not have any detectable health problems.

The antisense approach has already been demonstrated to work in the laboratory, and we will very likely see clinical trials with fusin antisense in the near future. The object will be to get the antisense gene into bone marrow stem cells, so that T cells can be produced for the rest of that individual's life that lack expression of fusin on their surface.

The "Poison Pill" Strategy

This approach would be used for cells already infected with HIV. The "poison pill" is a protein called **thymidine kinase (tk)**, produced from a gene taken from herpes-type viruses. Any cell in which tk is present can be killed by the drug acyclovir. Acyclovir is an FDA-approved drug routinely used to treat severe herpes infections and is completely safe in humans. Acyclovir is one of a very few drugs effective against a virus. The reason herpes viruses are sensitive to acyclovir is tied to the presence of the tk gene prod-uct. If we could selectively express the herpes tk gene in human T cells that are HIV infected, those cells could be killed by adminis-tering acyclovir.

There are several strategies for getting tk genes to function spe-cifically in CD4 T cells. One way is to place the tk gene under the control of a **long terminal repeat (LTR) promoter**. A promoter allows genes to be read inside a cell. The LTR promoter (taken from HIV) only functions in a human cell when the HIV tat gene product is also present. So genes under control of an LTR

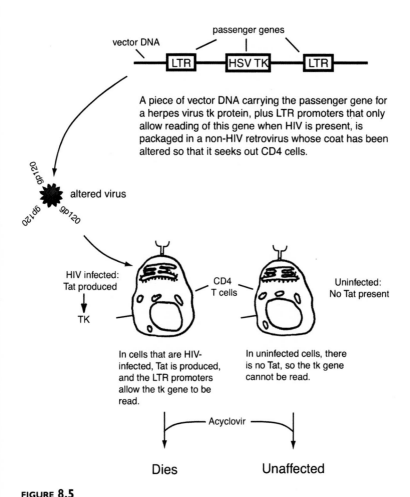

FIGURE 8.5

CD4 T cells in which the herpes tk gene is expressed under a tat promoter will be killed by acyclovir, depriving HIV of a place to replicate. Cells not infected by HIV are not affected by acyclovir.

promoter, like our tk gene, are *only* turned on, and their corresponding proteins produced, in cells that are infected by fully competent HIV.

If such a delivery vector were introduced into someone with an active HIV infection, the vector would deliver its genome plus any passenger genes into their CD4 cells. If any of those cells are actively infected with HIV, then tat produced by HIV will activate LTR and induce expression of the tk gene (Figure 8.5). If the cell is not HIV infected, tk will not be produced. The patient is then given acyclovir. HIV-infected CD4 cells die; uninfected CD4 cells are not affected.

Will We Ever Have an AIDS Vaccine?

Both of the approaches to managing HIV infection we just described will doubtless be effective and could potentially help bring the HIV/AIDS pandemic under control. Clinical trials for both (in one form or another) are currently under way or in the planning stage. But it is unlikely that either would be able to completely rid the body of HIV. Both would greatly reduce the amount of virus in the body and together with currently available drugs could have a major impact on patient survival.

And it is hard to see how they could have much of an impact on the spread of HIV in developing countries, where they are most needed. The technology and the necessary follow-up and compliance strictures are just not feasible in third-world countries, and even much of the second-world ones.

As with all infectious diseases, what is wanted ultimately—even in first-world countries—is a cheap, safe, stable, and effective vaccine. There is good reason to think an HIV vaccine *could* work; it's just that so far no one has been able to produce one that does. We think a vaccine could work because we already have vaccines in hand for the nonhuman primate form of HIV—the **simian immunodeficiency virus (SIV)**—which causes AIDS in monkeys, chimps, and other primates. We also have a vaccine for **feline**

immunodeficiency virus (FIV) in cats. And there are a number of sex workers in Africa who appear to have developed natural, effective immunity to HIV, indicating that the human immune system is capable of responding to at least some strains of HIV.

Part of the problem of making a human HIV vaccine is that it is considered too dangerous to use killed or attenuated virus as a vaccine. Both of these approaches are known to have a slight risk of containing incompletely neutralized virus. Particularly in terms of developing a prophylactic vaccine—one intended to prevent disease in healthy people—giving someone a vaccine with even the slightest possibility of inducing an almost uniformly lethal disease is unacceptable.

So vaccines to date have been based on purified fragments of, or even individual molecules isolated from, HIV. As you now know, these would be taken into antigen-presenting cells by endocytosis, be processed through the class II MHC pathway, and end up inducing almost entirely an antibody response. Any number of HIV-specific antibody vaccines have been produced, but these have proved ineffective. Antibodies can only see molecules on the surface of HIV, and these mutate too rapidly to make such vaccines of any use.

Natural infections by any virus produce both antibodies and CD8 T cells. We have known for some time that people who resist an HIV infection the longest (those with the longest latent period) are those in whom we see the strongest CD8 T-cell responses early in the infection. The 200 or so sex workers in Kenya who have developed natural resistance to HIV all have very high levels of CD8 killer cells specific for HIV. So the current emphasis is on CD8-inducing vaccines, and as we have seen, one of the best ways of doing that is with a DNA vaccine. Such a vaccine would be completely safe—it would not contain any material taken from an intact virus. It would stimulate preferentially CD8 cells. And it would be stable enough for use in developing countries.

A number of HIV DNA vaccine trials are now under way, in this country and around the world. The World Health Organization (WHO) has made HIV, together with tuberculosis and malaria, *the* major targets for vaccine production in the next decade. These three infectious diseases together account for the vast majority of infectious disease deaths worldwide. HIV DNA vaccines all target proteins within HIV that do not mutate, or that mutate very slowly. Virtually every such protein is now the target of a cDNA vaccine, and most such vaccines target several invariant proteins all at once (Figure 8.6, Table 8.2).

One major problem with developing a vaccine for a disease such as AIDS is the very long time between infection with HIV and the onset of disease, which averages 8 to 10 years. That means that people entered into a vaccine trial today won't yield any information about the effectiveness of the vaccine for at least that many years. Some people entered into these trials will be farther along in their infections when vaccinated, in the hope that the vaccination may help them resist onset of the disease. But it could well be 10 years before we can fully evaluate many of these vaccines.

We do not know whether these vaccines will achieve "sterilization" in vaccinated individuals—whether the virus can ever be completely eradicated from the body, even by a combination of vaccines and antiretroviral drugs. It may be that HIV will become

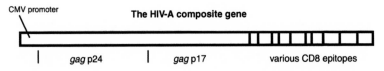

FIGURE 8.6
An HIV-specific DNA vaccine, designed to elicit CD8 killer T cells, for a clinical trial in Kenya. "Various CD8 epitopes" are gene fragments encoding portions of HIV proteins known to induce a CD8 T cell response (fragments of *nef, env,* and *pol*). The gag proteins p24 and p17 are known to induce strong CD8 cell responses. The composite gene has been inserted into a modified vaccinia virus that is not pathogenic in humans. It will be targeted to skin, a rich source of dendritic cells.

TABLE 8.2
Some HIV DNA Vaccines Currently in Clinical Trials Around the World

HIV ANTIGEN	VECTOR	WHERE?
env, gag, pol	pox virus	Thailand
gag	adenovirus	Worldwide
gag, pol, nef, env	naked DNA	Kenya
env, gag, RT, nef	pox virus	S. America
gag, RT, rev, tat, vpu, env	naked DNA	Thailand
gag, pol, nef, env	naked DNA	United Kingdom

Some of these vaccines are delivered as part of a genetically engineered virus; some rely on injection of the encoding DNA alone.

essentially an opportunistic pathogen, maintaining a balance between itself and the immune system. Perhaps occasional booster immunizations will be necessary. But from our experience with drug treatments alone over the past 20 years, it seems pretty clear that an effective vaccine may hold the only hope we have for a long-term solution to the AIDS crisis.

The importance of developing an AIDS vaccine has been recognized by the Bill and Melinda Gates Foundation, which has poured hundreds of millions of dollars into WHO-monitored efforts around the world in recent years. The U.S. National Institutes of Health will bolster this with a commitment of over $300 million, beginning in 2006. This marks a major shift from studying the virus itself, in hopes of developing effective drugs, to a study of the interaction of the virus with the immune system, in hopes of developing natural host immunity to the virus. It is our last, and perhaps best, hope.

9

When the Wall Comes Tumbling Down

PRIMARY IMMUNE DEFICIENCY DISEASES

One of the tried and true ways laboratory scientists use to determine the function of some part of a living organism—a cell, an organ, even a gene—is to disable it and wait to see what happens. That was what Bruce Glick did (Chapter 4) with the bursa of Fabricius, and by following his nose, he found that B cells and antibody could not be the only explanation of immunity. Faced with having to explain his results, others eventually stumbled onto the T-cell arm of the immune system.

A similar approach has been used to dissect out virtually every element of the innate and adaptive immune systems in laboratory animals. We can only do this in the lab—for all sorts of reasons—mostly because we can't do these kinds of experiments directly in humans.

But nature has in a sense done these experiments for us. The generation of all the cells and molecules involved in the human immune system is an incredibly complex process, and like all physiological processes, is directed by genes. For example, guiding the descendents of the hematopoietic stem cell found in bone marrow through all the steps necessary to become mature T or B cells, or any of the myriad other cells of the immune system, requires hundreds of different genes.

Genes also encode all the cytokines, antigen receptors, and even the enzymes used by phagocytic cells to digest the microbes they engulf. If any one of the genes needed to assemble the immune system does not function properly or is missing, the consequences can be disastrous. Depending on how early in the assembly sequence such gene defects occur, entire segments of the immune system can be wiped out. But even lesser defects can have a major effect on how the system functions, because all the parts are so delicately interconnected.

So when functional defects—mutations—creep into immune system genes, we are almost always left with an immune system that doesn't cover all the bases. The result is what we call **primary immune deficiency** diseases. They can affect innate immune responses, adaptive responses, or both at the same time (Table 9.1). As many as 50,000 people—mostly children—suffer from primary immune deficiencies at any given time in the United States, although many more likely have subclinical disorders they have learned to live with. The underlying mutations have arisen in germ-line genes—genes that are passed on from generation to genera-

TABLE **9.1**
Some Primary Human Immune Deficiency Diseases

Disease	Defect
X-linked agammaglobulinemia	Low antibody production
Common variable immune deficiency	Low antibody production, mostly IgA and IgG; IgM normal
Chronic granulomatous disease	Macrophages, neutrophils can't kill engulfed bacteria
Chediak-Higashi syndrome	Can't kill engulfed bacteria
Wiskott-Aldrich syndrome	Defective T cells
Hyper-IgM syndrome	Failure to produce IgA, IgD
Severe combined immune deficiency disease (SCID)	No functional T cells

tion. At least 100 such inherited diseases have been identified in the medical literature. In the sections that follow, we look at just a few of these.

X-LINKED (BRUTON-TYPE) AGAMMAGLOBULINEMIA (XLA)

XLA is caused by a faulty gene located on the X chromosome, so it affects only boys. One of the mother's X chromosomes carries a defective XLA gene. Among any sons produced, all would inherit the father's Y chromosome; half would inherit the mother's "good" X chromosome, and half the "bad." The XLA gene regulates the maturation of B cells, and thus controls the production of antibodies. XLA boys have very few antibodies in their system at birth and never develop the ability to make them throughout life. They are highly susceptible to many bacterial infections, which set in soon after birth. They have difficulty gaining weight and attaining normal height. Fortunately, they can be treated with monthly injections of pooled human gamma globulins, the fraction of blood that contains antibodies that other people have made to most common microbial antigens. Most bacterial infections, if they occur in spite of gamma globulin injections, can be managed with antibiotics.

COMMON VARIABLE IMMUNE DEFICIENCY (HYPOGAMMAGLOBULINEMIA; CVI)

This is an unusual primary immune deficiency in that, while genetic and heritable, it usually doesn't show up until the teen years. The exact gene involved has not been worked out, but this is also a B-cell disorder resulting in low antibody production, mostly of the IgG and IgA types. It affects both sexes equally and shows up as increasing bacterial infections. The treatment is essentially the same as for XLA.

CHRONIC GRANULOMATOUS DISEASE

This is a very rare disease affecting mostly males, although it is not strictly X-linked. Several genes are probably involved. It shows up soon after birth, when bacterial and fungus infections become increasingly common in these infants. Fevers, rashes, swollen lymph nodes, and even boils (bacterial abscesses) are common. The problem in this case is not in the production of antibodies, but in the ability of phagocytes to destroy antibody-tagged microbes after they have been engulfed. The failure to resolve the infection, however, causes continuous influx of macrophages to the infection site, and these soon form large, palpable masses called **granulomas**, which can begin to interfere with the function of nearby normal tissues. These children are also commonly anemic. Draining of abscesses, where necessary, and intensive treatment with antibiotics are required to resolve many infections that in most children would be relatively trivial. In some infants, a bone marrow transplant can correct the underlying defect by supporting the development of fully functional phagocytes.

CHEDIAK-HIGASHI SYNDROME

This is a complex disease that is not restricted to the immune system, but can disable some critical immune system functions. The gene defect in this case interferes with the release of granule-stored materials from all cells in the body, not just the immune system, so many systems of the body are affected directly or indirectly. In terms of immune function, neutrophils cannot release their granules, which contain important microbe-fighting materials. CD8 T cells and natural killer (NK) cells also cannot release the granules that contain perforin, involved in killing cells infected with viruses, bacteria, or parasites. Frequent infections make the short lives of afflicted infants increasingly miserable. The disease is untreatable, and the children only live a few years.

WISKOTT-ALDRICH SYNDROME (WAS)

This is another X-linked primary immune deficiency. The clinical manifestations of the disease include impaired T-cell function and a low platelet count. (Platelets are involved in blood clotting.) Patients usually present with frequent bleeding problems and infections. The gene that, when defective, causes this disease codes for a protein that is important in helping cells of the immune system react properly to activating signals. For example, T cells in WAS patients can recognize a peptide/class I MHC signal at a dendritic cell surface, but they cannot easily translate this signal into activation of the T cell. This same basic defect also impairs B-cell activation and NK cell function. Affected children can be managed by treating each of the clinical problems as they arise. In severe cases, a bone marrow transplant, to replace defective cells with normal ones, may be tried. As we will discuss below, this is a risky procedure generally used only when it is felt a patient is at great risk.

HYPER-IgM SYNDROME

Humans ordinarily produce five classes of antibody (Figure 2.1). The first antibody produced after B-cell activation is IgM. After B cells have produced a burst of IgM and settle down to become memory cells, they switch from IgM production to either IgG or IgA. Most antibody produced by humans comes from restimulated memory cells, so these latter two classes tend to predominate in normal human serum. This switch in immunoglobulin class is controlled by helper T cells. A protein called **CD40 ligand** (CD40L) in the membrane of CD4 cells interacts with the CD40 molecule on B cells to trigger this switch. There are several gene defects that can result in a lack of CD40L in T cells. In one form of this disease, restricted to males and accounting for nearly three-fourths of cases, the CD40L gene itself, located on the X chromosome, is defective.

Patients with this syndrome basically have no memory cells, and so cannot respond to infections as efficiently. IgG and IgA antibodies are also always more specific for antigen than IgM, and IgA plays a special role in gut immunity. Treatment usually involves gamma globulin injections to make up for this deficit, plus antibiotics as necessary. The CD40L molecule is also involved in other T-cell functions, which are likewise compromised.

COMING TRULY NAKED INTO THE WORLD: SCID

While all primary immune deficiencies can cause serious disease, with current access to pooled human gamma globulins and antibiotics, most of the B-cell deficiencies can be controlled. T-cell defects are a little more difficult to manage and may require a bone marrow transplant in the worst cases. But when an infant is born with neither T-cell nor B-cell function, the outcome in the vast majority of cases is death within the first year. That's what happens in a primary immune deficiency called **severe combined immune-deficiency disease (SCID)**.

SCID is truly as bad as it sounds. It is an inherited disease involving mutations in a single gene affecting T-cell function. Children born with this defect have no T-cell maturation from birth, and thus no B-cell function or CD8 killer cells. They have no adaptive immunity. They are basically tiny AIDS patients, and like AIDS patients are prey to virtually every pathogen, extrinsic or opportunistic, but in this case, from birth onward. The lack of defense against viral infections is usually apparent first. Some protection against viral disease may be conferred during the first year by innate immune mechanisms, and protection against some viruses and most bacteria may be provided by antibodies crossing the placenta from mother to child, or in breast milk. And the response to some bacteria is T-cell independent. But these infants will eventually be assaulted by round after round of T-dependent bacterial infections as well as viruses.

Keeping these children alive is a daunting task, for which only the best hospitals are equipped. Even then most do not survive. Too often they are brought to the hospital with advanced microbial infections that simply cannot be controlled in time to prevent death. Once SCID became generally recognized and management of it made a standard part of a pediatrician's training, it was realized that many of these infants developed fatal complications from immunizations with highly (but often not completely) attenuated vaccines shortly after birth. How many of these infants died from vaccinations prior to this realization is unknown. Fortunately, SCID is a rather rare condition. Only about 40 "SCID kids" are born each year in the United States.

As can be imagined, the outlook for infants with SCID is bleak. With the very best management and supportive care, they may survive the first two years of life; viral infections are the most common cause of death. Until recently, bone marrow transplants were the only hope for long-term survival. Bone marrow from a closely tissue-matched donor, with its stem cells, should be able to completely replenish all the cells of the immune system, including both T and B cells. Even then the outlook is poor; less than half of those so treated survive beyond a few years.

The best known case of a SCID kid was a young boy named David, who became known to millions of Americans and others around the world as the "Bubble Boy." David was born in 1971. Because a previous male child born to his parents had proved to have SCID and died from it a few months after birth, the risk of a second SCID child was known in advance. The form of the disease (there are at least a dozen different subtypes) in this family was what is known as **X-linked SCID** and affects only male children (Figure 9.1). David's parents decided to take the risk with another pregnancy. Amniocentesis at the fifth month of pregnancy showed that the child would be male; only males get this disease. At that time there was not yet a way to predict from amniocentesis whether a male fetus would have inherited the defective chromosome.

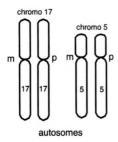

autosomes

The human genome is distributed over 23 chromosome pairs. 22 of these pairs are internally identical and are called "autosomal" chromosomes. One of each pair is inherited from the mother (m), and one from the father (p). Thus we all have two copies of every autosomal gene. If one bad copy is inherited, we can still function in most cases.

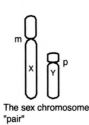

The sex chromosome "pair"

The sex chromosome "pair" is not really a pair. The x-chromosome is always inherited from the mother; the Y always from the father. Although the Y chromosome evolved from the X chromosome, the Y chromosome has very few genes - only those necessary for determination of the male sex. Thus almost none of the genes on the X chromosome have a corresponding gene on the Y. If a bad copy of an X-linked gene is inherited, its effect cannot be offset in males by a good copy on the Y chromosome. Thus only males get the disease.

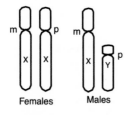

Females   Males

It follows that in X-linked disorders such as SCID or XLA, the defective gene was inherited from the mother, who is said to be a "carrier". She is asymptomatic.

FIGURE 9.1
X-linked genetic diseases.

David was delivered by cesarian section and transferred within seconds to a sterile incubator until his immune status could be determined. It was soon obvious that he, too, carried the defective gene causing SCID. The major rationale for this highly unusual approach to managing a SCID child was the hope that if he could be kept alive long enough by keeping deadly pathogens away from his body, either a suitable bone marrow donor could be found or his immune system might somehow establish itself and allow him to fend for himself. Since virtually nothing was known about the basis of this disease in the early 1970s, these were not unreasonable hopes.

Young David quickly became the longest living SCID patient, untreated except for sterile isolation. He was repeatedly tested for signs of T- or B-cell responsiveness; there were none. When he reached the toddler stage of development, he was moved into a sterile tent that allowed him to crawl and eventually stand. The tests continued—still no response. When he began to walk and run, the tent became the "bubble," a complex system of interconnecting plastic tubes that allowed considerable freedom of motion, within obvious limits. He was adored by his nurses as well as his family and given plenty of physical contact and cuddling through sterile gloves reaching into his bubble.

NASA even built a small spacesuit for him when he was 6 so he could be taken into the outside world as well. He outgrew it within a year. A sterile transporter was also developed so that he could be taken home and develop a sense of place with respect to a family. He was given the basics of an education in his bubble and at home; his nurses and tutors found him a bright, somewhat mischievous youngster, virtually indistinguishable from other boys his age. But his immune system never developed; the only thing between him and certain death was a few millimeters of plastic sheeting and high-quality air filters.

As David continued to grow and develop, it became clear that something simply had to be done. He was healthy and vigorous and at 12 years of age showing the first signs of normal sexual maturation. He had not yet begun showing outward indications of a curiosity about sexual matters, but clearly his situation was approaching a critical stage. No one had really thought this far ahead; no untreated SCID kid had ever lived this long. His medical team found themselves in an ethical dilemma of gigantic proportions, with no guidelines for how to proceed. The prospect of maintaining him any longer in a sterile bubble—for how long? 10 years? 20? 50?—was becoming increasingly untenable. How do you talk with a child like this about the future, a concept he now understood only too well?

Finally, in a joint decision between his doctors and his parents, it was decided to give David a bone marrow transplant, with his sister (now 15) as the donor. In most cases, marrow from sibling donors has a higher chance of successful acceptance than marrow from a complete stranger. However, David and his sister were not particularly tissue compatible, which is one reason a bone marrow transplant was not attempted earlier.

Nevertheless, it was decided to proceed. Marrow was removed from David's sister and treated to remove mature T cells. Mature T cells are not found in bone marrow per se; they are a contaminant from the harvesting procedure, when blood vessels woven throughout the marrow are broken, allowing mature blood cells to mingle with the precious bone marrow stem cells. It was thought at the time that mature T cells contaminating donor marrow might be responsible for a major barrier to successful transplantation— graft versus host (GVH) disease, which can be lethal.

In GVH disease, mature donor T cells in the incoming marrow, being fully competent immunologically, regard the new host as a gigantic transplant, which they immediately set about trying to reject (Figure 9.2). The graft, in effect, is rejecting the recipient. This can be fatal in a quarter to a third of patients receiving a bone marrow transplant, which is why such transplants are carried out only in the most serious situations.

When his sister's bone marrow was ready, David was taken from his bubble in a sterile transporter to a sterile operating room and infused with the marrow. He was kept in a sterile postoperative recovery room, and then returned to his bubble. For the next several weeks, everything seemed to go well. But then he developed symptoms that seemed possibly related to GVH disease: weight loss, gradually increasing fever, vomiting and diarrhea, and abdominal tenderness. Appropriate steps to control GVH were immediately undertaken, but he did not respond. There were also signs of a viral infection. His condition grew rapidly worse; he

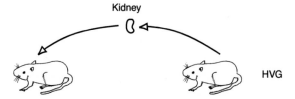

Figure 9.2 In a typical host-vs-graft reaction, the T cells in mouse A (the host) recognize MHC antigens on the mouse B kidney (the graft) as foreign, and reject it.

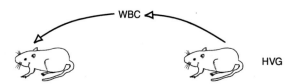

When the graft from mouse B consists of white blood cells, the same thing occurs, and the white cells are also rejected

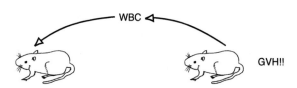

FIGURE 9.2
Graft versus host disease.

finally died on the 124th day posttransplant, in February 1984. He was 12 years old.

It turned out David did not die from GVH disease after all. In fact, his sister's bone marrow never engrafted into his own bone marrow, which frequently happens. His sister, like much of the population, harbored a virus in her B cells called Epstein-Barr virus (EBV), which in her immunologically compromised brother jumped over to his B cells, inducing a fatal lymphoma.

Before David died, some of his T cells were harvested and frozen away for future study. In 1993, scientists working with DNA isolated from his cells were able to pinpoint the defective gene that causes his kind of SCID. The gene encodes a receptor on developing T cells that receive a crucial cytokine from the thymus environment. The cytokine is there, in the thymus, but the developing T cells can't "see" it, and they never mature into functional T cells.

A SECOND FORM OF SCID AND
A RADICAL NEW TREATMENT

There is another form of SCID, equally as deadly as the X-SCID that afflicted David, called **ADA-SCID**. About half of all SCID cases seen clinically in the United States are X-SCID; ADA-SCID accounts for roughly a quarter. (Ten other variants of SCID account for the rest.) Although ADA-SCID results in exactly the same disease symptoms that David experienced, the underlying genetic defect is completely different. ADA-SCID is caused by a mutation in the gene that codes for an enzyme called **adenosine deaminase (ADA)**, which is involved in the metabolism of DNA. The gene for ADA is not X-linked, but rather is found on one of the other chromosome pairs. Development of ADA-SCID requires the inheritance of *two* defective gene copies, one from each parent, and it affects newborn boys and girls equally.

ADA is an enzyme used in all cells of the body, and someone born with two defective copies of the ADA gene has no ADA in any cell of the body. This leads to a build-up of a chemical called *deoxyadenosine*. This is not a problem for most cells, but T cells convert the deoxyadenosine into another chemical that is exceedingly toxic, and the cells die. Children born with no ADA thus lose all their T cells and suffer exactly the same set of consequences as children with X-SCID.

Bone marrow transplantation can also correct this form of SCID, but as we have seen this is a risky procedure. An alternate treatment for ADA was developed, in which patients were given injections of the ADA enzyme itself, in a form known as PEG-ADA. Enough of it may get into the developing T cells to rescue them before they die. But in the case of a 4-year-old girl named Ashanti DaSilva, even this form of therapy had not worked. By September of 1990, the T cells in her blood had fallen to 50 per cubic milliliter of blood, a level ordinarily seen only in patients in the terminal stages of AIDS. As in David's case, there was no well-matched sibling to act as a donor, so a bone marrow transplant was considered very risky. Her family and doctors were running out of options. But this young lady would be offered an option never before given any human being: she would be invited to become the very first human gene therapy patient.

In this first-ever gene therapy trial, it was decided to introduce a normal, healthy copy of the ADA gene directly into her T cells. A few of her scarce T cells were withdrawn from her blood and placed into culture dishes under conditions that would cause them to grow and expand, and they were exposed during this growth period to literally billions of copies of the healthy gene. These gene copies were delivered in a retrovirus, which as we have seen inserts itself into the host cell DNA. The retrovirus is disabled so that it cannot replicate itself, but the human **passenger gene** it is carrying will be produced in the cell the virus infects. After a week and a half of growth and exposure to the retrovirus and its passenger, her modified T cells were placed back into her bloodstream through a simple intravenous drip.

And so one of the monumental moments of medical science passed very quietly—no heroic surgery, no lights or cameras, just a cloudy suspension of cells dripping for a couple of hours into the forearm of a rather shy 4-year-old girl. She was awake and alert during the entire procedure. After the infusion, Ashi, as she was called, was monitored closely for any signs of an adverse reaction; there were none. As a precaution, she has been kept on PEG-ADA throughout the trial. After another injection a month later, the level of healthy T cells in her blood began to climb. Some of these cells were taken back to the lab and examined for their expression of the healthy ADA. The new gene had indeed found its way into the DNA of her T cells and was rescuing them—and her—from death.

Most importantly, little Ashi began to show signs of being able to make antibodies, proving that the repaired T cells were able to carry out one of their key functions—helping B cells. She also showed signs of having acquired T-cell functions associated with the ability to kill virally infected cells. In less than a year, her T cell count was 1,250, within the normal range for healthy humans. A second young girl with ADA-SCID was treated by exactly the same procedure just a few months after Ashi. Both of these young ladies now have vastly improved T-cell function and more vigorous and healthy immune systems generally. They attend school and are very active in, among other things, the national March of Dimes campaign to raise funds for childhood diseases. To date, several dozen young children have been treated for ADA-SCID by gene therapy.

And so the medical revolution that may change the face of medicine in the twenty-first century was actually born in the final decade of the twentieth. The first use of antibodies in the treatment of infectious disease happened at almost the same point in the nineteenth century.

GENE THERAPY FOR X-LINKED SCID

Three years after the gene in which mutations cause X-SCID was identified in David's DNA, it was isolated and cloned. By this time gene therapy for ADA-SCID had been working successfully for several years, so normal copies of the human X-SCID gene were scaled up as quickly as possible for use in similar trials, using the same strategy as that in ADA-SCID. Good copies of the gene were inserted as passengers into a disabled retrovirus vector and used in gene therapy for X-SCID by a French team beginning in 2000.

By this time, it also had become possible to insert a passenger gene into a sample of the patient's bone marrow stem cells, which is far preferable to inserting it into mature T cells. The reasoning goes like this. We know that if a bone marrow transplant is accepted and does not cause GVH disease, it is possible to cure SCID. The new, normal bone marrow produces normal, healthy T cells, and that is all that is necessary. We know that if we remove bone marrow from someone and then put it back into that same person, those bone marrow cells will find their way back to the marrow and function perfectly normally. So, while we have someone's own bone marrow out of his or her body, why not use that opportunity to insert a normal copy of the defective gene into the bone marrow cells before putting them back in? Once the defect is corrected in the bone marrow stem cells, those cells will continue to produce normal T cells for the lifetime of the patient.

The French team eventually performed this procedure on 11 male infants. The underlying defect was corrected in all of them, and all recovered their T-cell immunity. This was tremendously exciting, since it showed that genes can indeed be delivered to bone marrow stem cells and function as intended.

However, three of the infants treated by the French researchers for X-SCID subsequently developed leukemia, and it is now clear that this was due to the treatment. In general, these leukemias are treatable, but one child died. The trials were of course immediately suspended, and other centers around the world that were about to begin using this technique put their plans on hold. This is another major ethical dilemma. These infants would all die if untreated. If some small number of them dies as a result of treatment—as is the case with bone marrow transplants, by the way—which ethical path do we follow? And who makes that choice?

The decision was made, and accepted internationally, that for now gene therapy will be used to treat X-SCID only in those cases where a closely matched bone marrow donor is not available. But the parents of the child involved will have to be thoroughly informed of the risks of the procedure and will have to participate fully in the decision to begin treatment.

We now know that the type of retroviral construct used in the French study tends to insert in the DNA near genes that can cause cancer and activates them. Why this has not shown up in treatment of ADA-SCID patients, most of which now also treat bone marrow, is unclear. In fact, the same types of retroviral construct have been used in literally hundreds of other gene therapy trials, for other genetic diseases, without incident. The reasons for this discrepancy must be fully understood.

But the setback in the French trials is, in the end, a technical problem, with numerous possible solutions. The underlying scientific rationale for using gene therapy for X-SCID is very strong. The focus now has been shifted to the laboratory, where each of the corrective solutions can be explored and rigorously tested to ensure the problem has truly been taken care of. In the meantime, the limitations on its use, which are not unreasonable, will remain in effect.

Gene therapy clinical trials for correction of chronic granulomatous disease and Wiskott-Aldrich syndrome using gene therapy are under way; trials for hyper-IgM syndrome are in the planning stages. Although stumbling blocks have been encountered along the way, it seems very likely that mortality from inherited diseases of blood cells will soon become a thing of the past. As discussed in the last chapter, this form of treatment may also be useful in treating AIDS; the technology is identical in both cases, and as we can see from this chapter, it is already working.

Note also that in doing gene therapy of this type we are not tinkering with the human genome. The changes brought about in an afflicted individual do not affect his or her sperm or ova, and so are not passed on to the next generation. They die with the individual, hopefully at the end of a long and disease-free life. This is an important point in this age of possible human cloning and other technological horrors.

10

When the Immune System Is the Problem, Not the Solution

HYPERSENSITIVITY AND ALLERGY

The immune system truly is a wall that stands between us and the universe of microbial agents of disease. The slightest crack, or missing stone, can spell disaster. So the possibility that this beautiful, elegant defense system could also cause harm was at first very difficult to accept. Some of the earliest findings pointed in that direction, but were largely ignored.

But as scientists looked ever more closely at the immune system during the first half of the twentieth century, that possibility became first a probability, and ultimately a reality. It seemed in a number of situations, at least in the laboratory, that exposure to a foreign antigen did not always lead to protective immunity, but rather to a state in which subsequent exposure to the same antigen could elicit a violent, often harmful, and occasionally fatal syndrome. This phenomenon of overreaction became known as *hypersensitivity*. It would eventually be recognized for exactly what it is: a defense system out of control. But this realization took time.

The discovery of hypersensitivity began with the work of two young French scientists, Paul Portier and Charles Richet, at the very beginning of the twentieth century. They were studying the immune response in dogs to toxins produced by sea creatures, including that of the sea anemone. As they began their anemone studies, they found to their great surprise that after an initial injection of toxin, rather than developing a protective immune response, the dogs became **hypersensitive** to subsequent doses of the toxin. The following quote from their laboratory notebook illustrates what they observed:

10 Feb [1902]—26 days after first injection—the dog was in perfect health, cheerful, active; the coat was shiny. On this day at 2 PM it was injected with 0.12 cc toxin per kg. Immediately produced vomiting, defecation, trembling of front legs. The dog fell on the side, lost consciousness, and in one-half hour was dead.

This finding was completely unexpected, but completely reproducible. Initially, it was thought to be due to the action of the toxin itself, that is, an alteration in the animal's tissues caused by the toxin that made the animal unusually sensitive to a subsequent dose of the same toxin. A year later, however, it was shown by another French scientist that the same result could be obtained by immunizing an animal with normal, nontoxic proteins as well. As with the toxin experiments, hypersensitivity was not always induced, but it could be if the initial immunization regimen was properly managed.

The experiments with normal protein antigens stunned everyone working in the new field of immunology and had to be repeated numerous times by many different researchers before they were fully accepted. Portier and Richet called this phenomenon **anaphylaxis**, from the Greek meaning essentially "opposite of protection," This seminal work eventually led to a Nobel Prize for Richet in 1913.

Clearly a new hypothesis of immunity was required. Since hypersensitivity responses to both toxic and nontoxic substances were found to be antigen specific (i.e., hypersensitivity in each case was restricted only to the immunizing antigen and no other), it was difficult to escape the conclusion that they were immunological in nature. The problem was that no one wanted to believe that a physiological system designed to protect—the immune system—could also maim, and even kill. In these early days of immunology, when no one could be quite sure what it was they were discovering and how it would fit into the overall picture of immune protection, it was easy to set such unsettling observations aside. But in fact they never did go away, and would ultimately prove to be harbingers of a darker side of the immune system barely imagined at the beginning of this new medical science.

One of the most dramatic examples of hypersensitivity can be seen in the guinea pig. Guinea pigs are unusually susceptible to the induction of hypersensitivity. The injection of extremely small levels of things as innocuous as egg white protein can make guinea pigs hypersensitive. Subsequent injection of the same antigen will induce a state of **anaphylactic shock**. Within minutes of the second dose, the animal becomes restless, begins rubbing its nose and eyes, and experiences difficulty in breathing. Its fur stands on end; it may urinate and defecate. It hiccoughs violently, gasping for air. Blood pressure begins dropping rapidly, as does temperature, and the heartbeat becomes markedly irregular. Lowered blood pressure deprives the brain of oxygen, leading to disorientation and loss of muscular control. If the reaction is strong enough and the animal is not rescued with drugs, convulsions set in and death from asphyxia follows shortly. Postmortem examination shows lungs completely stretched out of shape and filled with fluid and air. Of particular importance, the openings from the windpipe to the lungs are almost completely closed off due to contraction of the surrounding muscle. As gruesome as all these facts may seem, mark them well, for we will see shortly that the identical set of symptoms can develop in people.

As late as 1927, eminent authorities continued to declare that the hypersensitivity reactions observed in laboratory animals did not occur in humans. They were eminently wrong. The human immune system is indeed capable of turning its weapons inward against its human host. As we will see, sometimes the result is no more life threatening than a mild (or even severe) case of allergy. The violent response to manipulated forms of antigen described earlier for guinea pigs are indeed rare in nature, for humans as well as animals. But the following two cases show that on occasion natural hypersensitivity reactions in humans can be quite severe.

Case 1. A 25-year-old white male laboratory employee reported for work with a sore throat and was given 500,000 units of penicillin by injection into the left deltoid muscle. Although there was no history of having been treated with penicillin previously, he became apprehensive within two to three minutes and complained of burning and tingling sensations of the scalp, "tightening" of the chest and throat, respiratory distress, and headache. He began to perspire profusely, rapidly developed edema about the eyes, mouth, and throat, and collapsed. Cyanosis was marked and pulse could not be felt, nor blood pressure ascertained. A tracheotomy was performed, artificial respiration was applied, and aminophylline (0.5 mg) and epinephrine (1 mL 1:1,000) were given intracardially. Oxygen was administered through a catheter. The pulse became manifest, the blood pressure could be determined, and breathing resumed. The patient remained unconscious for 12 hours. Subsequent recovery was uneventful. A history revealed that he had been working in a tissue culture laboratory handling media containing antibiotics, including penicillin, and had suffered allergic attacks during the pollen season. He was instructed to wear a dog tag thereafter inscribed with the warning that he was dangerously allergic to penicillin.

Although reported in the typically dry and detached clinical style favored by medical scientists, there is no doubt that the situation being described was drama of the highest order. The classical signs of anaphylactic shock were fortunately spotted immediately by an alert and experienced physician, who did not hesitate to take extreme action on the spot: cutting open of the windpipe to aid breathing, forced introduction of air together with pure oxygen, and direct injection into the heart of stimulants to revive and maintain an effective pulse. After a period of unconsciousness, the patient underwent an "uneventful" recovery and was sent on his way with a reminder to wear a MedAlert bracelet. The next patient was not so fortunate.

> *Case 2.* A 30-year-old white farm laborer suffering from allergic asthma and hay fever was admitted to the allergy clinic. On the first occasion he was scratch-tested on the arms with a number of pollens prevalent in the area, with negative results. The next day, tests for sensitivity to various foods were conducted on the skin of the back. After several tests had been performed, the patient suddenly complained of difficulty in breathing and collapsed. Despite the administration of aminophylline and atropine sulfate intravenously and epinephrine intracardially, together with artificial respiration, the patient expired within 15 minutes. Autopsy revealed visceral congestion, edema of trachea and epiglottis, subpleural hemorrhages over the right lobe, and marked emphysema of both lungs.

The autopsy findings with this unfortunate patient are not markedly different than for the guinea pigs described earlier, in which hypersensitivity was deliberately induced in the laboratory. The cause of death in both cases was asphyxiation due to constricted air passage to, but mostly from, the lungs. Although the provoking antigen was not defined for patient number two, most likely one of the test antigens related to foodstuffs previously ingested by the patient stimulated the abrupt and violent response that led to his death. We will talk about food allergies (normally a very mild form of hypersensitivity) later in this chapter. But it is worth noting that both of these patients had previous allergic disorders. This is the usual finding in cases of severe hypersensitivity problems.

ALLERGY IN HUMANS: THE TIP OF AN ICEBERG

Roughly one in five Americans suffers from one form or another of **allergy**. The most common form of allergy is based on an immune reaction called **immediate hypersensitivity**. Immediate hypersensitivity reactions are called that because the symptoms manifest themselves within minutes of exposure to antigen in sensitized individuals and peak within a few hours.

The list of substances that provoke immediate hypersensitivity responses in humans is virtually endless and may well include almost anything in the biological or chemical environment. Of course, no one person (fortunately) ever develops immediate hypersensitivity to all possible provoking antigens (called **allergens** when we are specifically talking about antigens that induce allergic reactions). While some allergens may induce immediate hypersensitivity in large numbers of individuals—certain plant pollens; animal dander; house dust—others are as individual as people are: specific drugs or chemicals; a particular brand of makeup; certain foods. The list of symptoms is similarly long: runny noses, itchy eyes, shortness of breath, rashes and eczema, diarrhea, and so on. It's little wonder it took many years before it was determined that all of these various problems and symptoms are related by a common mechanism, let alone that they are all caused by the body's own immune system.

That immediate hypersensitivity reactions in humans are caused by antibodies was suggested by experiments—in part carried out on each other—by two German physicians, Carl Prausnitz and Heinz Küstner. Küstner was allergic to a protein in cooked fish; Prausnitz wasn't. In addition to experiencing distressing symptoms when he ate cooked (but not raw) fish, Küstner also found that when he injected tiny amounts of a protein extract of cooked fish into the skin of his forearm, a rapid and marked reaction ensued. In about 10 minutes a small welt began to arise at the site of injection. It looked very much like a mosquito bite and itched like one. The welt grew rapidly until it was as much as an inch and a half in diameter and was surrounded by a red, patchy region up to four inches across. After about 20 minutes, Küstner began to experience the more generalized, systemic manifestations of a classical hypersensitivity reaction: the itching spread to other parts of his body, he began to cough, and he had difficulty breathing. After another 20 minutes the symptoms leveled off and then slowly drifted back to normal.

The critical part of the experiment involved his colleague Prausnitz, who was not sensitive to fish in any form. When Prausnitz was injected under the skin with the cooked fish extract, absolutely nothing happened, no matter how much was injected or how often. But if Prausnitz was first injected with some of Küstner's serum and a short time later injected under the skin with fish extract, the exact pattern of local swelling and itchiness seen in Küstner developed in Prausnitz.

This experiment demonstrated in the clearest possible way that the agent active causing immediate hypersensitivity in humans *circulates in the blood*. The skin test developed by Prausnitz and Küstner provided a way of routinely screening for allergy to specific substances in humans; the "P-K" test has been a standard of the allergy clinic for many years.

The antibody responsible for hypersensitivity reactions is called IgE. IgE is one of five major classes of antibodies made by humans (Figure 2.1). B cells specializing in IgE production tend not to hang out in lymph nodes and spleen, but are found in the skin, lungs, and intestinal lining—the points of entry for many pathogens. For some reason, a few individuals seem preferentially to make IgE-type antibodies in response to certain environmental antigens. The first time someone makes IgE antibodies, nothing much happens. For example, an initial bee sting may result in nothing more than the discomfort of the sting itself. But a subsequent sting from the same type of bee may result in a mild or severe hypersensitivity reaction. Why some people make IgE in response to particular antigens and some don't is not well understood.

The reactions that lead to hypersensitivity are due to the unique homing properties of the tail portion of IgE. The initial exposure to allergen triggers the production of IgE. When the IgE antibodies build up to a critical level, they begin to bind to two special immune cells, **mast cells** and **basophils**. These cells have **Fc receptors** on their surface that specifically bind IgE, just as macrophages and neutrophils bind to IgM or IgG antibody tagging bacteria or viruses.

Mast cells and basophils are filled with granules that contain a variety of highly active pharmacological reagents, chief among which is **histamine**. When antigen (allergen) comes into the system a second time and interacts with this surface-bound form of IgE, the basophils and mast cells are triggered instantly to release the contents of their granules, including histamine, into the bloodstream. It is this degranulation reaction that leads to many of the unpleasant side effects associated with immediate hypersensitivity and allergy.

We know a lot about histamine, and it is clear that together with a few other biochemical components of mast cells and basophils, histamine can account for virtually all of the phenomena associated with immediate hypersensitivity reactions. When histamine binds to blood capillaries, it causes them to enlarge and become more permeable to blood fluids. This is responsible for the rash associated with allergic reactions that take place at the body surface. Of greater concern is the fact that the increased permeability of blood vessels, if it occurs systemically (throughout the body), will also cause a drop in blood pressure and lead to a state of potentially lethal shock.

Another problem caused by histamine is that it binds to the smooth muscle surrounding the bronchioles leading into the air sacs of the lungs, causing them to contract. This leads to a marked constriction of the passageways for air into and out of the lungs. One of the highest concentrations of mast cells in the body is found in the lungs. When histamine is released from mast cells into lung tissue, the resultant constriction of bronchioles becomes a major factor in the respiratory distress, and even respiratory failure, accompanying immediate hypersensitivity reactions. Histamine also triggers mucus-secreting cells to spill more mucus into the airways, further impeding airflow. Air can usually be forced into the lungs by strong, voluntary contractions of the diaphragm (gasping), but subsequent relaxation of the lungs is not strong enough to force the air back out. In the experiments described earlier on anaphylaxis in guinea pigs, autopsy usually showed distended lungs that floated in water. Asphyxiation occurs with the lungs full of stale, used air.

SPECIFIC FORMS OF HUMAN ALLERGY

Hay Fever

Descriptions of what is almost certainly hay fever **(allergic rhinitis)** date almost as far back as the beginning of written history. But despite its name, hay fever is not a fever, and only rarely is it caused by hay! It is most often caused by pollens or other plant-associated products carried by wind; allergies truly can be due to "something in the air." In North America, one of the most serious offenders is **ragweed**, a plant that spreads its pollen throughout much of the summer and early autumn.

But allergic rhinitis can be caused by any airborne allergen—chemicals, dust, microbial spores, animal dander, fibers, or insect parts—in addition to pollen. As the term allergic **rhinitis** implies, the nose is a particularly sensitive target. The nose is unusually rich in small blood vessels and secretory glands, related to its role in warming and moistening incoming air. Even in the absence of an allergic reaction, the nose may secrete as much as a quart of water every 24 hours as it moistens the air passing by. Hairs in the nasal passages help trap airborne particles and are thus a natural filter for incoming allergens.

The nasal passages are also lined with IgE-secreting B cells and with both mast cells and basophils, so the allergic response in the nose is both rapid and rabid. Histamine release from mast cells causes local blood vessels to dilate and become more permeable. Fluids cross out of the blood vessels into surrounding tissue spaces, creating a sense of swelling and pressure. These fluids need to escape from the area, and in part are secreted through glands and membranes, leading to an endlessly runny nose. The reaction rarely remains confined to the nose, however, and usually involves the roof of the mouth and throat (contributing to the annoying sensation of postnasal drip) and particularly the eyes, the surrounding tissues of which have their own IgE-producing B cells and mast cells.

Hay fever–like symptoms caused by other than seasonally produced plant or animal products will of course be with the poor sufferer year round. The term hay fever is usually applied to seasonal allergic manifestations, with the more general term **perennial allergic rhinitis** reserved for year-round upper respiratory tract allergies. Interestingly, perennial rhinitis affects females much more than males: the ratio is about three to one. The most common allergens in perennial allergic rhinitis are substances like animal dander or fur, airborne molds or spores, house dust (usually contaminated with dried insect parts), minute fibers from cloth, and anything else floating around in the air that is capable of calling up IgE antibodies in sensitive individuals.

Drug and Venom Anaphylaxis

The allergens associated with "hay fever" generally induce symptoms that are annoying but hardly life threatening. On the other hand, a few substances can induce hypersensitivity reactions that are every bit as violent as those described earlier for laboratory guinea pigs. Among the more common allergic reactions that can result in anaphylactic shock are those to certain drugs—particularly penicillin and its derivatives—and reactions to venomous bites, particularly by insects such as bees and certain biting ants. Almost everyone has heard of someone who went into shock and nearly died as the result of a bee sting or as the result of an unsuspected allergic reaction to penicillin.

Like hay fever, these reactions are mediated by IgE and mast cells. The reactions are swift and, if not rapidly treated, deadly. The symptoms begin within minutes of exposure to the allergen and may be accompanied by a range of symptoms—faintness, breathing difficulties, nausea, and tingling of the skin and scalp. Extreme breathing difficulties and a drop in blood pressure are the most life-threatening symptoms and require immediate treatment. As with other forms of allergy, these symptoms do not occur on the first exposure to the allergen. The initial exposure simply builds up high levels of drug-specific IgE, which take up residence on the surfaces of mast cells and basophils. Subsequent exposure, particularly if the allergen is introduced into the bloodstream, provokes the anaphylactic response from these IgE-primed cells. In the United States, there are still several hundred deaths each year from anaphylactic shock in response to drugs or venoms. The formal cause of death in such cases is usually asphyxiation or the complications of vascular collapse and shock.

Asthma

If an inhaled allergen penetrates beyond the nose–throat area into the lungs, the more serious problem of asthma may arise. Asthma can be caused by the very same allergens that cause hay fever. In fact, the older literature refers routinely to "hay asthma." Asthma occurs in all known human populations; in its various forms it probably affects about 2% of the people in the United States. Although the management of asthma has improved dramatically in the past 50 years, it is still responsible for some 2,000 to 3,000 deaths per year, mostly among the very young and the very old.

Asthma is a complex medical condition that has causes other than immediate hypersensitivity (e.g., other than an overactive immune response). The forms of asthma caused by inhaled allergens interacting with IgE on mast cells are referred to as **extrinsic asthma**, because like hay fever they depend on interaction with substances that enter the body from the outside. But essentially the same symptoms can be caused by a variety of other factors *not* involving allergens or the immune system. Stress, for example, can be a major inducer of asthmatic attacks in certain individuals. Some of the neurotransmitters released during emotional or traumatic stress can either trigger or certainly exacerbate an allergic attack of any kind, but particularly asthma. Exercise is also a well-known potentiator, and possibly an inducer, of asthma in sensitive individuals. Clinicians refer to these kinds of asthmatic attack as **intrinsic asthma**. In fact, many asthma attacks are a combination of the two forms, making treatment a real test of the physician's skill.

To a considerable extent, extrinsic asthma, like allergic rhinitis, is caused by the release of histamine and other mediators from mast cells and basophils. Certainly the early stages of an asthmatic attack are closely dependent on IgE and mast cell levels. However, other elements of the immune system are also involved in asthma, making even the immunological aspects of this disease more complex. Shortly after an IgE-mediated asthmatic attack begins, the lungs are invaded by white blood cells, which stimulate formation of thick mucus that gets secreted into the bronchioles, together with the excess fluid accumulating in response to histamine.

The severe difficulty in breathing during an asthmatic attack is thus the result of several related pathologies. Histamine and another cytokine called **leukotriene** cause constriction of the bronchioles, narrowing the passageways for air into and out of the lungs. The accumulation of mucous secretions caused by white cell infiltration and the build-up of fluid in the bronchioles further impedes the flow of air. Finally, histamine acting locally in the lungs leads to the accumulation of fluid in regions where the lungs normally take up oxygen from inhaled air, leading to oxygen depletion in the blood.

True extrinsic (immune-based) asthma is more common in children than in adults. Up to 10% of preteen children may experience asthma to some degree, usually along with other allergies such as hay fever or drug hypersensitivity. Asthma can be a terrifying experience for both parents and children. In serious attacks, with widespread constriction of the bronchioles, it becomes difficult for the child to expel air from the lungs, and an asthma attack can begin to approximate the anaphylactic shock syndromes described earlier in guinea pigs. Although the lungs may be full of air, not enough oxygen is getting into the bloodstream. Yet the brain tells the lungs to try to take in more air! The result is severe gasping and wheezing. Fortunately, a wide array of highly effective bronchodilators is available at any pharmacy. Often as the child gets older, both asthma and related allergies decrease substantially.

Because the symptoms are not trivial, asthma can be an expensive disease in terms of medications, doctor visits, and, in adults, time off from work. Generating some 30 million visits to the doctor per year and several billion dollars in treatment costs, asthma is clearly a mainstay of both the medical and pharmaceutical industries in this country.

Food Allergies

What could be more central to staying alive and healthy than eating? Considering that most of us will eat 25 or 30 tons of food in a lifetime, the likelihood of an adverse response to at least some foods should be pretty fair. Food allergies certainly have the potential to be among the most life threatening, or at least health threatening, of all the immediate hypersensitivities. Although this is a possibility only rarely realized, scores of people die of anaphylactic responses to food allergens each year in the United States. Fortunately, the vast majority of food allergies lead only to nausea, vomiting, cramps, and diarrhea, and may also involve distress outside the gastrointestinal tract such as itching, hives, or asthma. This is not particularly pleasant, but it's also not particularly life threatening and is easy to avoid once the offending foodstuff is identified.

A distinction should be made (although it often isn't!) between **food intolerance** and **food allergy**. The latter is a true immunological hypersensitivity to a particular food; the former includes basically everything else that causes a problem with that food. Numerous studies have shown that the majority of self-diagnosed "food allergies" are simply gastrointestinal distress identified in the patient's mind with a particular food eaten around the time the distress occurred. Usually less than one-third of these self-reported allergies holds up with controlled testing in the allergy clinic using the suspected food allergen. The prevalence of true food allergy in the general population is actually around 2% or less, and nearly all of this is in children. Food allergies in adults are more rare, but as we saw in an earlier case history, they can be very deadly indeed.

Almost all food allergies are to proteins. As with all food, serious digestion of protein begins in the stomach. When partially digested protein passes from the stomach to the small intestine, it is hit with an infusion of powerful protein-degrading enzymes from the pancreas that continue the digestive process. In a normal, healthy adult, this process will be essentially complete; that is, the proteins taken in as food will be completely degraded into amino acids, and these amino acids will be transported across the intestines and into the bloodstream for use as building blocks in the synthesis of new proteins. So there is really nothing for the immune system to react to.

Occasionally, however, small amounts of partially digested protein may cross the intestine. In persons with gastrointestinal disorders such as ulcers, some food proteins may even cross the gut without being digested at all. Once proteins or protein fragments large enough to be antigenic cross into the bloodstream, they are no different from any other foreign protein entering the blood and have the potential to induce an immune response.

The reason food allergies are more common in children, particularly during the first two to three years of life, is that at birth the human digestive system is still somewhat imperfect. There is less acid in the stomach, and lower levels of digestive enzymes. Many of the barriers to intact proteins crossing out of the intestines are not yet fully developed. Maturation of an infant's digestive system is aided by breast milk, and breast milk also brings in antibodies to help neutralize potentially antigenic substances. In most cases allergic symptoms simply disappear with time, but occasionally they will persist into adulthood.

The most common sources of food proteins causing immediate hypersensitivity reactions in humans are milk, egg whites, peanuts, fish, and soy, more or less in that order. Allergies to peanuts and other nuts in particular can be quite deadly. While the allergy is to a protein associated with peanuts, traces of this protein may be found in peanut oil, and foods cooked in peanut oil can trigger a violent allergic response. This points to one of the real difficulties in tracing food allergies. One oft-quoted case describes a violent reaction in someone who had just eaten a tuna sandwich. The reaction was not at all to something in the tuna itself; the knife used to prepare the sandwich had just been used to cut a peanut butter sandwich. The poor patient nearly died!

Food additives or preservatives may also be allergenic. As with all other allergies, allergic symptoms develop only after a second or third exposure to the offending allergen. The symptoms may show up in almost any part of the body, with the digestive tract being only one of them. The underlying mechanisms in food allergy are exactly the same as in any other allergy: selective production of IgE in response to a particular food allergen entering the bloodstream, and then interaction of that IgE with mast cells. Because both the IgE and the allergen are free to travel anywhere in the body, food hypersensitivity can manifest itself in many different forms: hives, asthma, or fatigue as well as cramps, nausea, or diarrhea.

The most effective treatment for food allergy (or food intolerance, for that matter) is avoidance of the offending food. For a few unfortunate human infants, cow's milk can be a potent allergen; switching to either breast milk or formula usually solves the problem. The most important thing is to have the condition properly diagnosed by a doctor or an allergist. If the problem is truly a food allergy, it is a good idea to have this confirmed and recorded as part of a child's permanent medical record, since it may indicate a general predisposition to allergy.

WHY HYPERSENSITIVITY?

Why do we have hypersensitivity? What possible good can it do? What is its relation to positive, protective immunity? We don't really know in every case.

Classical (IgE-mediated) allergic responses are the hardest to rationalize, for one simple reason: we do not know why IgE exists in the first place. The body has four other classes of antibody; why does it need IgE? There is reasonably good evidence that in some parasitic infections, IgE is selectively produced and may take part in clearing out the parasites. But other elements of the immune system are also called into play in these infections, and it is far from clear that IgE is critical to the host response even in cases where it is induced. Moreover, during infections with parasites, it is not just parasite-specific IgE that is elevated, but IgE in general. So it is not obvious that the induction of IgE during parasitic infections is antigen-specific. In those rare individuals with a deficiency in IgE (including a complete absence of IgE), there are no detectable immunological problems with parasites or any other pathogens. Detailed studies of IgE production in vivo suggest that there is a fairly sophisticated regulatory apparatus for preventing the production of IgE. That is rather bizarre; why have a class of antibody whose production you try to prevent? We don't do this with any of the other classes of antibody.

It is possible that some living things in our environment may have co-opted our IgE system for their own protection. Our extremely strong reaction to bee stings and to certain plant products may reflect the fact that they have evolved toxins that are very efficient in inducing an aggressive (and highly unpleasant) IgE response in us. That would certainly encourage us to keep a close eye on what's around us and avoid those organisms causing such grief!

So is IgE nothing more than a fossil image of a dangerous episode in the evolutionary history of humans, or a reflection of manipulative organisms in our environment? We simply don't know. For that matter, why do we need mast cells? There are also other cells that carry out many of the functions of mast cells. One rarely if ever hears of immune deficiency diseases in which IgE or mast cells are selectively missing. Is this an indication that they are relatively unimportant in the overall scheme of immunity, such that when they are deficient or missing altogether, we never even notice the difference?

Unquestionably, a good many people still die each year from IgE-mediated anaphylactic shock. Before we understood anaphylaxis and learned how to treat it (thanks largely to work in animals), doubtless more people died. But the numbers were probably never very large, certainly not on the order of those dying at the time from diphtheria, smallpox, or the plague. And again, these are the pathogens that the immune system evolved to protect us against. Failure to respond promptly and forcefully to such pathogens means certain death for an unprotected individual. The immune system we ended up with has IgE as part of its repertoire. We don't know why it's there, or what good it does, but there it is. Current thinking among immunologists is that IgE-mediated allergies may just be the price we pay as a species for an immune system that otherwise does an outstanding job of keeping us alive in a dangerous world.

The Immune System and Cancer

WHAT IS CANCER, AND HOW DO WE GET IT?

Just a few decades ago, our thinking about cancer, and especially its treatment, was as fragmented as the disease itself appeared to be. Oncologists were a frustrated lot; there seemed to be as many different diseases called cancer as there were different cells in the body; any one of them could become cancerous, and each of the resulting diseases seemed to require a completely different approach to treatment.

Our thinking about cancer has changed remarkably in recent years. Bone cancer still looks different from brain cancer; skin cancer is still treated differently than lung cancer. But the current focus is on understanding what causes cancer in the first place, and here the emphasis is on what cancers have in common, rather than on what makes them different. Cancer can be contributed to by external agents—radiation, chemicals, and some viruses—or by mistakes made within a cell when it reproduces its DNA. And indeed, any cell or tissue in the body *can* give rise to a tumor. But ultimately *every cancer is a disorder of DNA*. All cancer cells share one single, common feature: they have their lost ability to regulate DNA synthesis and cell division. And the regulatory elements governing these processes lie in the DNA itself—in our genes.

Cells lose control of DNA replication and cell division through mutation or loss of the genes that keep these processes tightly regulated in normal cells. One category of such genes consists of **oncogenes**, whose normal protein products are involved in telling a resting cell when to divide and then guiding it through the process. Ordinarily this only comes about in response to a **growth signal** arriving at the cell surface from somewhere else in the body, telling the cell to start dividing. When this external signal is withdrawn, the oncogenes turn off, and the cell promptly returns to the resting state. Mutations in oncogenes lead to the turn-on of cell division in the absence of normally required signals. The result can be unscheduled cell division and initiation of a tumor.

The second category of genes involved in cancer is **tumor suppressor genes**. These genes monitor DNA damage. When cells divide rapidly and continuously, they can make a lot of mistakes copying the DNA handed down to daughter cells. Damaged DNA is considered by the cell as extremely dangerous. Tumor suppressor gene products attempt to repair damaged DNA, but if the damage cannot be repaired, the suppressor genes instruct the cell to fall on its sword—to commit apoptotic suicide.

So the development of a tumor requires the mutation of several genes. There must be at least one mutation in a gene activating cell division—an oncogene—that starts the cell down the pathway toward proliferation in the absence of an appropriate signal. But there must also be a mutation in one or more of the genes whose specific purpose is to detect DNA damage that inevitably accompanies such events—tumor suppressor genes. There is also a gene encoding an enzyme called **telomerase**, which in most normal cells is turned off. Telomerase is needed to assure the integrity of DNA in dividing cells and must be turned back on if a tumor is to grow continuously. This requires yet another mutation.

The need to accumulate multiple mutations is one reason most cancers do not arise until relatively late in life. The likelihood of getting just those three mutations in the same cell is pretty small. But we have a lot of cells, and eventually this can—and does—happen. On the other hand, some cell types may be more susceptible than others. Increasingly, it seems that many, perhaps most, tumors arise within specialized cells found in all tissues called **stem cells**. These are the cells that constantly replenish dead or damaged cells within our tissues and organs. Stem cells are in a sense primed to enter cell division more easily than most other cells in our body, and it may take less to get them to cross over to a cancerous state. This provides another reason for vigorously pursuing research into stem cell function.

CANCER AND THE IMMUNE SYSTEM

What if the immune system could respond to tumors, could recognize them as somehow aberrant, like a virally infected cell or a transplant? What if those tumors that do develop in us are just the rare renegades that manage somehow to slip past our immune defenses? Would that mean we could then somehow pump up the immune system to increase its "search and destroy" function as a means of combating cancer?

These questions were a major driving force in the early days of cancer. Adding fuel to the fire were observations that when scientists tried to pass tumors from one animal to another, they were rejected. But the idea that this had anything to do with the tumor nature of the transplant was laid to rest in the 1930s, when it was shown tumor rejection in these cases was no different than rejection of a normal kidney or a piece of skin. Tumors were rejected because they were from another person, not because they were tumors—end of story. **Tumor immunology** per se was pushed onto the back burner for nearly 20 years.

Interest in immune responses to tumors was rekindled in the 1950s, when people began studying cancers induced by chemicals. For example, the highly carcinogenic chemical methylcholanthrene, when painted on the skin of a mouse, will cause a sarcoma almost every time. These experiments were done in **inbred strains** of mice. Inbred strains are strains produced by repeated brother–sister matings. After 20 or so generations, all the members of the same sex of an inbred strain are essentially genetically identical twins. And just as with human identical twins, they can exchange cells, tissues, and organs with no possibility of immune rejection.

Transplanting tumors or anything else between members of the same inbred strain is like transplanting a piece of skin from your right arm to your left.

What these researchers found was that if a chemically induced sarcoma was surgically removed from one mouse of an inbred strain and small pieces of it transplanted to a mouse of the same sex and inbred strain, they would grow into a large tumor and eventually kill the mouse. But if a piece of the tumor was reimplanted back into the mouse of origin, it did not grow! However, this mouse was *not* resistant to other tumors—the reaction was absolutely specific.

This showed beyond a doubt that tumors *do* have distinct antigens, and that mice can indeed respond immunologically to a tumor. Obviously this immunity is not sufficient to prevent a primary tumor from popping up in the first place. But by immunizing a mouse, by allowing the tumor to grow for a limited time before removing it surgically, mice could be made much more resistant to future attempts to implant the same tumor. The animal had developed antigen-specific immunological memory.

But as with so many other situations prior to the discovery of T cells, these researchers were stumped. When presumably tumor-immune serum was transfused into a naïve mouse of the same strain, which was then given a fragment of the tumor that had induced the antiserum, the naïve mouse was not protected: the tumor grew just as well as it did in a mouse receiving no serum at all. So it seemed that antibodies—the only immune defense known at the time—could not be the basis of tumor immunity. Eventually it was shown that in tumor immunity, as in delayed-type hypersensitivity (DTH) reactions, effective immune responses could be transferred from one animal to another with cells, but not with antibody.

By the 1970s, everyone was convinced that T cells must be the major immune defense against cancer. In fact, it was proposed that tumor surveillance might be the major raison d'être for killer T cells. We now know that, unlike immune responses to intracellular parasites—where antibodies can play a major role in host defense, since most microbial parasites are exposed to extracellular fluids at some point in their life history—the effective immune response to cancer is indeed almost entirely cellular. Antibodies play a minor role at best. Both CD4 and CD8 T cells play major roles in compromising tumor survival.

NK CELLS AND CANCER

But there was increasing evidence that T cells might be only part of the story. For one thing, mutant mice that lacked a thymus, and thus T cells, had only a slightly increased incidence of spontaneous tumors. Most worrying of all was the question of specificity. In every reaction in which T cells are involved, they exhibit exquisite specificity toward foreign antigen (albeit in conjunction with MHC).

That is not what was always seen when researchers first looked at killing of tumor cells in the laboratory. It was frequently found that when white blood cells were taken from cancer patients and tested against their own cancer cells in a culture dish, the cancer cells were killed. This seemed very exciting. But some researchers, wanting to be as precise as possible, used white cells taken from noncancer patients as controls. To their dismay, lymphocytes from non–tumor-bearing individuals often displayed as much or more cytotoxicity toward tumor targets as did lymphocytes from those bearing tumors.

While at first dismissed as "artifacts," it soon became apparent this was the rule, not the exception. A monumental study published in 1973, looking at antitumor "killer cells" from 995 cancer patients and white cells from 1,099 non–tumor-bearing controls, showed no obvious statistical difference between the two groups in their ability to kill a random panel of tumor cells in the lab.

Needless to say, these findings were not enthusiastically received by those searching for a role for "classical"—highly antigen-specific—killer T cells in tumor immunity. But where some saw only disaster for their field, others saw an opportunity to discover something new, and it was quickly established that normal healthy individuals do indeed have a subpopulation of white cells that, without any prior sensitization, will recognize, attack, and destroy at least some tumor target cells.

These effector cells became known as **natural killer (NK)** cells. They were assigned to innate immunity: the collection of infectious disease–resistant mechanisms that are genetically imprinted in each organism and are fully functional at birth, independent of contact with environmental antigen. But interestingly, NK cells use exactly the same killing mechanisms CD8 T cells use.

As we discussed in Chapter 5, when talking about the involvement of NK cells in defense against viruses, NK cells selectively kill cells that have lost expression of class I MHC on their surface. How does this happen in tumor cells? Well, some tumors, primarily leukemias and other white blood cells cancers, for some reason seem to lose, or at least greatly reduce, class I expression as part of the tumor forming process for reasons that aren't always clear.

But NK cells are active against more than just blood cell tumors. How do cells lose their MHC? Most likely, through mutation. Development of a detectable tumor from a single aberrant cell requires thousands, if not millions, of cell divisions. Each time all the DNA in a cell (the entire **genome**) is copied during cell division, mistakes are made. These are called **copy errors**. There is "editing" machinery in each cell to correct these errors, but a great many still get through. The best guess at present is that during all this wild cell division and copying and editing of DNA, lots of mutations will creep in.

Some of these mutations may well result in loss, or at least reduction, of class I MHC. This is a great advantage for the tumor cell in which it happens. Lack of class I MHC means that the cell cannot present its tumor antigens to CD8 T cells, and that the tumor cell and its descendants have just escaped from a very important host defense against tumor cell growth! These tumor variants will enjoy an obvious growth advantage over other cells in the emerging tumor that still express class I and get picked off by CD8 cells, and they quickly become the dominant tumor cell type.

That's where NK cells come in. When NK cells recognize these kinds of tumor cells, they are not recognizing tumor cells per se, but absence of class I MHC. Many tumor cells do not lose their class I, and these are ignored by NK cells. Any cell losing its class I MHC may escape being killed by CD8 T cells, but it now becomes susceptible to killing by NK cells. So NK cells may be a primary immune defense against many white blood cell tumors, which seem to lose their class I MHC very early in tumor development, and a secondary defense, playing backup for CD8 cells, for other forms of cancer.

BACK TO CD8 T CELLS

CD8 cells could kill tumors by two possible means. One would be by direct killing of the tumor cells, using the perforin and Fas mechanisms described in Chapter 5. But CD8 cells also secrete interferon-γ (IFN-γ), and this cytokine inhibits a process that is absolutely essential for a growing tumor—**angiogenesis**. Tumor cells divide constantly, and they need enormous amounts of nutrients and oxygen to support their growth. This requires a rich supply of blood vessels. If tumor cells do not get adequate food and oxygen, they die within a day or so. IFN-γ can block development of this blood vessel network and starve a tumor to death. This is likely to be as important as direct tumor cell killing in CD8 control of tumors.

So just what is it that CD8 cells see on the surface of a tumor that makes them want to kill it? They can only see something connected to a class I MHC protein, of course, so it has to be a peptide that is being made inside the cell. But what kinds of MHC-bound antigenic peptides tell them that a cancer cell is not normal and may pose a threat to the host?

Over 70 **tumor antigens** have been identified in humans, and they seem to fall into three categories.

Some are derived from proteins normally expressed at low levels in a limited number of cells, but which are greatly overexpressed in tumor cells. Perhaps this breaks the delicate state of tolerance and allows an immune response to develop against a self protein; no one is completely sure.

Some antigenic peptides are derived from proteins normally present only at restricted stages of embryonic development but expressed—again, often at very high levels—in tumor cells. Carcinoembryonic antigen (CEA), abundantly present in fetal gut but barely detectable in adults, is a common antigen in colon cancer. α-Fetoprotein, a major serum protein produced by fetal liver but essentially absent in adults, is a common antigen in human liver cancer.

One of the most promising categories is the so-called **tumor-unique antigens**. As we saw above, a cancer cell becomes cancerous in the first place through mutations in genes that regulate the ability of a cell to divide and that prevent it from dividing when it shouldn't. Any protein in our bodies that mutates is potentially a foreign protein from the point of view of the immune system. Oncogenes and tumor suppressor genes are examples of this category and are prime targets for cancer vaccines (see below).

SO IF THE IMMUNE SYSTEM DETECTS CANCER, WHY DO WE STILL GET IT?

That's a question scientists have been wrestling with for the last half-century. By definition, a tumor that becomes detectable has escaped every immune defense we have thrown at it. How does it do that?

There are several hypotheses to explain this, but they remain just that—hypotheses. Still, they are reasonable hypotheses and form the basis for a good deal of research into cancer treatment.

One thought is that some tumors have levels of tumor antigens too low to be detected by the immune system, or tumor antigens that only weakly stimulate the immune system. This might be particularly true of tumor-unique antigens, where the mutations in an oncogene or a tumor suppressor gene may be too subtle to be detected by the immune system.

It may also be that some tumors secrete substances that are immunosuppressive. We know that many cancers do make people less able to respond immunologically to many things. But what the nature of these immunosuppressive substances might be is anyone's guess. Doctors who do organ transplants might really like to know!

Another possibility is that either the tumor antigen is not being processed inside the tumor cell for proper presentation at the cell surface by MHC or the level of MHC at the cell surface is too low to present tumor antigens effectively, but high enough to keep NK cells suppressed.

And, of course, the true explanation may not have been thought of yet.

IMMUNOTHERAPY AND GENE THERAPY FOR CANCER

But let's forge ahead anyway. How can we explore what we know about cancer, and the immune response to it, to our advantage, as early researchers had hoped a hundred years ago? A number of approaches have been tried over the years. For example, we've learned a lot about the various cytokines that regulate immune responses, and some of these have been used as cancer-fighting drugs in an attempt to boost the body's response to tumors.

In a few cases this does seem to have worked. The interferons—IFN-α, IFN-β, or IFN-γ—are now part of standard treatment for many tumors. Their exact mode of action is not always clear. They may act by promoting inflammation, which is likely harmful to rapidly growing tumors. But also, each of these interferons increases the level of expression of class I MHC on the surface of all cells, and this could enhance the effectiveness of CD8 killing of tumor cells with reduced class I.

A clever gene therapy variant of this idea has recently shown some startling results. Tumor cells are removed from a patient, irradiated, and then transfected with one or more genes encoding a variety of activity-boosting cytokines. The transfected tumor cells are then placed back into the residual tumor mass. Any CD8 cell that comes nosing around the tumor, possibly recognizing that something is not right but not able to mount an effective attack, will be bathed in a flood of activity-hyping cytokines pouring out of the tumor.

There is a certain perverse satisfaction in enlisting one's own tumor to encourage the immune system to destroy it. And it works. In a recent study with non–small cell lung cancer patients, one such tumor-embedded cytokine, called **GM-CSF**, caused very significant increases in survival among patients who had failed all other treatments.

A form of the passive immunity transfer technique we saw in chapter 2 has been adapted for tumor immunotherapy. We can use the patient's own T cells in something called **adoptive immunotherapy**. The evidence that CD8 cells provide a potent tumor defense is strong, and it seems likely that in many, perhaps most, cases, a more vigorous CD8 attack could turn the tide of battle in favor of the patient.

In adoptive immunotherapy, CD8 killer cells are isolated from a patient, expanded greatly by growing them under favorable conditions in the laboratory, and then transferred back into the patient's bloodstream. Partially depleting the patient's overall lymphocyte population by mild chemotherapy prior to transfer sometimes increases the effect of the transferred T cells, perhaps giving them more room to expand. Increasingly, CD4 helper T cells are also transferred in and seem to improve the CD8 killer cell effect.

An excellent source for tumor-specific CD4 and CD8 cells is the tumor itself. Tumors that can be removed surgically are dissected and dissociated into smaller fragments, and **tumor-infiltrating lymphocytes (TILs)** are washed out and recovered. These cells are already highly activated and can be grown in culture if supplied with appropriate cytokines that act as growth factors. If we know the tumor peptide that is most favored by CD8 cells, these can be fed to dendritic cells, which are cocultured with the T cells, resulting in selective expansion of the corresponding CD8 cell populations.

In many cases individual CD8 cells can even be isolated and **cloned**. Individual CD8 cells are placed into culture dishes, stimulated, and fed cytokines and their progeny expanded and kept as a separate line in the laboratory. All the members of such lines are identical clones of each other, resulting in populations of CD8 cells that are all highly specific for exactly the same tumor. In practice, several dozen of these lines are generated for a given tumor and screened for the most potent killers to transfer back into the patient. And since they all came from the patient, there is no possibility of rejection.

In a number of cases, when these TIL clones were infused back into the patient, they appeared to home in on the tumor and cause significant destruction, as measured by tumor shrinkage. The problem arose, however, that these TIL lines had become so dependent on cytokine growth factors in the laboratory that cytokines had to be infused directly into the patient's bloodstream together with the TIL. The levels of cytokines needed caused unacceptable side effects, and the procedure, although promising, had to be suspended for a while. But more recently, very potent CD8 TIL lines and clones have been generated that are much less dependent on cytokines, and these may be more suitable for immunotherapy.

The results look highly promising. Using adoptive immunotherapy, significant regression of tumors has been seen in malignant melanoma, certain leukemias, and a subset of Hodgkin's lymphoma patients. And perhaps more exciting, in many cases it appears that the transferred T cells have developed into memory cells, suggesting the possibility of long-term protection for treated individuals.

CANCER VACCINES

Producing a vaccine against cancer has been a dream since the earliest days of immunology. The goal of a cancer vaccine, however, is different than for an infectious disease vaccine. The number of different types of cancer is huge—as we said at the beginning, as many as there are different types of cells in the body. So undertaking mass vaccination programs prophylactically—before disease develops—is not going to be practical. Rather, we will want to develop individually tailored vaccines that can reverse the disease once it has become clinically detectable.

It is only in recent years that we have understood both vaccination and cancer well enough to make informed attempts at producing a cancer vaccine. Early attempts often involved something as simple as surgically removing a patient's tumor, grinding it up, irradiating it, and reimplanting it into the body along with adjuvants known to stimulate antibody responses in general. Occasional positive effects were seen, but they were not consistent and this approach was soon abandoned.

But now we realize that, as with vaccines for intracellular parasites, we need a vaccine that will selectively stimulate production of CD8 T cells, and for that we need to know which peptides associated with the cancer are most likely to produce an effective CD8 response. This approach has been tested in mice and has given impressive results.

In humans, a lot of work has been done to find out which peptides would be best to use for a variety of common tumor vaccines. The most impressive results have been obtained with malignant melanoma (Figure 11.1), and clinical trials with vaccines for this cancer have been under way for the past several years. Melanoma is one of the most difficult cancers to treat, particularly if it has spread beyond the original site. Once a tumor has spread, or become metastatic, only systemic treatments such as chemotherapy, and possibly a vaccine, can be effective.

Typical of the antigens defined for melanoma are MART-1 and

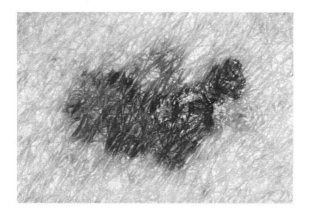

FIGURE 11.1
Malignant melanoma lesion.

gp 100. Both of these are nonmutated proteins normally found at low levels in skin pigmentation cells (melanocytes) but over-produced in melanocytes that have become cancerous. The genes for MART-1 and gp 100 have been cloned, and corresponding peptides have been identified that would make the best vaccines. Clinical trials using a gp 100 peptide, administered together with stimulatory cytokines, yielded tumor regressions in 42% of patients with advanced melanoma. In a second trial a selected peptide from MART-1 was shown to induce significant CD8 responses in melanoma patients, which correlated with a prolonged time to relapse. The melanoma trials have now advanced to include patients with less advanced disease, where the results can be ex-pected to be even better.

Vaccine trials with peptides from other cancers including breast, cervical, and pancreatic are under way and have given broadly similar results where reported. As with the melanoma trials, these trials are in the early stages and are restricted to patients with advanced cancer who have failed conventional treatments. How-ever, it must be admitted that at present, although CD8 responses have been good, tumor regressions have not been as impressive as with melanoma. But the melanoma studies have been going on much longer. Everyone expects that as we learn more about exactly which peptides produce the best CD8 responses in these other cancers, the success rate will improve.

So will DNA vaccination replace any of the current standard treatments for cancer patients? Certainly not in the immediate future. The first line of standard cancer treatment, where possible, is simple physical reduction of the tumor mass by surgery. In some cases that is a complete cure in itself. But for the majority of cases, where the tumor is inaccessible to surgery or where surgery reduces but does not completely eliminate the tumor load, follow-up treatment with other modalities is required. Radiation therapy can be used to clean up residual tumor at the primary tumor site. Chemotherapy can be used for the same purpose, and can also chase down and (hopefully) eliminate metastases that have spread beyond the primary site.

No vaccine, however effective, would likely displace surgery for accessible tumors. Reducing the tumor burden will make the job of any vaccine that much simpler. And to the extent that radiation or chemotherapy can reduce the tumor load further, the immunotherapist is just that much further ahead. Certainly, if a particular vaccine turns out to be particularly effective, it is possible that radiation therapy or chemotherapy could be reduced, or even possibly eliminated, given the serious side effects that come with these treatments. Given the recent successes of cancer vaccines and the tremendous opportunities that lie ahead for this new modality, it would be surprising indeed if we did not see it moving into the clinic, as part of the oncologist's bag of standard anti-cancer weapons, by the end of this decade.

12

Autoimmunity

IMMUNOLOGICAL TOLERANCE

Here's one of the central questions in immunology: how does the immune system know what is self and what is foreign? The molecules of which human "self" is made are basically the same as those used in the construction of any other biological organism, including pathogenic microbes. Clearly the immune system must be able to make this distinction, or else we would self-destruct. So how does the immune system avoid making antibodies or T cells reactive to self?

One of the earliest insights into these questions was made by Ray Owen in 1945. Owen was studying an interesting type of twin in cattle called a **freemartin**. Freemartins are genetically distinct ("fraternal") twins.[1] Normally, fraternal twins each have their own placenta, isolating them from each other while in utero. But freemartin fraternal twins share a common placenta. The fact that they are connected by blood during fetal life means that as embryos they share everything that moves around in the bloodstream. The cells of the blood mix freely prior to birth, and after birth each twin has a mixture of two genetically different types of blood. Even the stem cells from which all blood cells derive mix between the twins. As a result, this state of mixed blood types persists for life.

---

1. Freemartin twins also occur in humans, although they are extremely rare in natural births. They are less rare in births generated through in vitro fertilization, however. Human freemartins share the same characteristics as those described here for cattle.

This was a classic case of an extremely important point staring one right in the face; look a little to either side, and you'd miss it. But Ray Owen didn't. He thought about his freemartins and realized that if they had each had their own placenta and their bloods had *not* mixed before birth, then as adults they would definitely be intolerant of each other's blood. This is what we see in human fraternal twins, each with their own placentas: unless they happen by chance to have exactly the same blood type, it is no more possible to exchange blood between two fraternal twins than it is to exchange blood between any two randomly selected individuals, unless they too accidentally have the same blood type.

But freemartin twins are completely tolerant of each other's blood, no matter how genetically disparate they are, for life. They can also exchange other cells and tissues with relative impunity. This led to what would become one of the most important theoretical principles of immunology. Anything we are exposed to *prior to birth* will be regarded as self. But if we are exposed to the very same things *after birth*, they may be considered foreign.

This has been shown to be true in a great many laboratory experiments since Owen first reported his observations with freemartins. In mice and rats, as it turns out, the period during which prenatal tolerance can be induced actually lasts until one or two days after birth, making such experiments relatively easy to perform. For example, a newborn mouse injected with cells from an adult rat can, as an adult mouse, accept a skin graft from the same type of rat with no sign whatever of rejection. The rat skin and the accompanying fur, even if of a different color than the mouse's own, will last for life. The same piece of rat skin placed on an untreated mouse would be rejected almost instantly.

At or near the time of birth, the newborn animal (or the almost-born, depending on the species) takes one last look around, and basically says: "Okay, this is it; this is me. Anything other than this that I see from now on is foreign, is potentially harmful, and must be eliminated." This decision is communicated to the animal's immune system, which imprints it onto the T and B cells that are charged with making self/nonself determinations.

But since T and B cells live only a few weeks before they die and are replaced, each succeeding generation of T and B cells produced, for the rest of an organism's life, will have to learn the same information over and over again, without making a single error. If this process is perturbed in any way, the result may be autoimmune disease—not an accidental spillover of damage in the course of trying to remove a cryptic pathogen, but a genuine, unprovoked aggression against perfectly normal, healthy self cells.

Newly emerging T and B cells learn a lot about what is self and what is not from the thymic and bone marrow environments, respectively, in which they mature. It is estimated that 50% of B cells are eliminated in the bone marrow on the basis of potential self-reactivity. At least 90% of T cells are eliminated in the thymus for the same reason. But not all the information about self can be found in these environments, and thus many T and B cells with potential self-reactivity arrive in the body with this self-reactivity intact, but under various forms of control. Interruption of these controls is thought to be a common source of autoimmunity.

AUTOIMMUNE DISEASE IN HUMANS

Autoimmune diseases affect about 1 in 15 people in the United States. A partial list of some of the more common human autoimmune disorders is shown in Table 12.1. Almost every organ and tissue in the body can be a target for autoimmune disease.

There are a number of generalizations about autoimmune disease that seem to hold true. Although there are some relatively organ-specific autoimmune diseases, in fact almost all autoimmune diseases affect more than one system in the body. The relatively restricted diseases are just that—*relatively* restricted. Patients with

TABLE 12.1

Some Representative Human Autoimmune Diseases.

Ankylosing spondylitis[a]	Myasthenia gravis[a]
Autoimmune hemolytic anemia[a]	Pemphigus vulgaris
Autoimmune hepatitis[b]	Pernicious anemia
Autoimmune inner ear disease	Polychondritis
Autoimmune lymphoproliferative syndrome[b]	Polymyositis[b]
	Primary biliary cirrhosis[a]
Autoimmune thrombocytopenic purpura[a]	Psoriasis[a]
	Raynaud's syndrome
Bullous pemphigus	Reiter's syndrome
Cardiomyopathy[a]	Rheumatic fever[a]
Crohn's disease[a]	Rheumatoid arthritis[b]
Diabetes (type I)	Sarcoidosis
Degos' disease	Scleroderma[b]
Dermatomyositis	Sjögren's syndrome[b]
Fibromyalgia	Stiff-man syndrome
Goodpasture's syndrome[a]	Systemic lupus erythematosus
Grave's disease[a]	Ulcerative colitis
Hashimoto's thyroiditis[a]	Uveitis
Idiopathic pulmonary fibrosis	Vasculitis
Ménière's disease	Vitiligo[a]
Multiple sclerosis[b]	Wegener's granulomatosis[b]

[a] Relatively tissue restricted.
[b] Relatively non–tissue restricted. (Unmarked: intermediate.)

insulin-dependent (Type 1) diabetes, for example, almost always have other autoimmune problems. The spectrum of diabetes-associated autoimmune diseases (pernicious anemia, Grave's disease, Hashimoto's thyroiditis, to name just a few) is so broad that sometimes it's easier to think of diabetes as just one part of a broad-spectrum "pan-autoimmunity" that happens in a particular individual to affect the pancreas more than other organs.

Those diseases that almost always affect many different tissues in the body, such as lupus (systemic lupus erythematosus [SLE]), Sjögren's syndrome, or rheumatoid arthritis, have one peculiar feature in common: they tend to affect women much more than men. Whereas hemolytic anemia affects men and women more or less equally, arthritis is two or three times more frequent in women; lupus, 6 to 10 times more. Even some of the relatively tissue-restricted autoimmune diseases, such as myasthenia gravis (which we will talk about shortly), affect predominantly women. But, strangely, Type I diabetes, like hemolytic anemia, is an exception to this rule; it too affects men and women equally.

In autoimmune diseases affecting women more strongly, the disease appears relatively early in life, usually during the child-bearing years. There has been speculation that women are more prone than men to develop autoimmune disease because they have developed more powerful immune systems to protect their fetuses. Whatever the reason, it is clear that such autoimmune diseases are regulated by sex hormones. Studies in strains of mice in which the females spontaneously develop a lupus-like disease have shown that manipulating sex hormones can drastically alter the disease. In humans, males born with an extra X chromosome (XXY; Klinefelter's syndrome) have more lupus-type autoimmune disease.

In addition to the gender bias, most autoimmune disorders appear to have a genetic basis, in that they tend to "run in families." But the genetic link is only partial. In studies of genetically identical twins, only about a third would both have multiple sclerosis; half might both have diabetes; a quarter could both develop SLE. At autopsy—in the case of accidental death, for example—the apparently healthy twin often shows subclinical signs of the disease, suggesting that the genetic linkage is stronger than it appears from clinical diagnoses. And finally, as we will discuss later, there is very definitely an interplay between the mind and the immune system in autoimmunity. So these are very complicated conditions, indeed. Just talk to the 7% or so of Americans who suffer from them!

Most autoimmune damage is caused by low-grade, chronic inflammation, driven by both B cells and T cells, CD4 as well as CD8. It is very similar to the immunopathology seen in unresolved infectious diseases or immunopathologies. On one level, this makes perfectly good sense. It's like graft versus host (GVH) disease, where foreign, immunocompetent T cells are transplanted into an immunoincompetent person: the T cells find themselves surrounded by a gigantic foreign graft, and they begin to reject it. In autoimmune disease, our own T cells suddenly find themselves surrounded by a seemingly endless universe of foreign material. Is it a transplant? Is it a microbial infection? No, it is us. But the immune system sets about mounting exactly the same types of reactions as if we were our own transplant, or a sea of microbes. For reasons that are not clear, autoimmune damage, although quite serious in some cases, is only rarely fatal. But it can be very miserable, indeed.

In order to get a feeling for the range of disorders with an autoimmune basis, let's take a brief tour of a few of the major human autoimmune diseases.

AUTOIMMUNE HEPATITIS

Autoimmune hepatitis occurs about eight times more frequently in women than in men, and is found almost exclusively in women of northern European descent. The symptoms are essentially the same as in viral hepatitis: fatigue, weakness, jaundice, and dark urine. In addition, young women with this disease usually have disturbances with their menstrual cycles. The disease results when, for some unknown reason, the immune system begins to regard certain liver proteins as foreign and T cells begin to attack and destroy the liver cells. Antibodies are also formed to liver cells, as well as to muscle and even kidney tissue. If not treated properly, autoimmune hepatitis can progress into exactly the same kind of cirrhosis seen in viral hepatitis (chapter 6) and can be fatal.

Although similar to the viral form of hepatitis caused by the hepatitis B virus, even the most sensitive tests fail to detect any trace of active viral infection. And true autoimmune hepatitis is almost always accompanied by other autoimmune symptoms such as thyroiditis, arthritis, or myasthenia gravis, which does not happen in viral hepatitis. Autoimmune hepatitis responds well to corticosteroids, whereas this drug has minimal impact on viral hepatitis. But these differences can be fairly subtle, and it takes an alert and well-trained physician to make the proper diagnosis. It was many years before an autoimmune form of hepatitis, developing in the apparent complete absence of any extrinsic pathogen, was recognized and accepted for what it is.

This is a perfect example of why it was so difficult for both scientists and physicians to believe that autoimmune diseases are really, truly autoimmune, and not an attack on cells harboring faint traces of some hard-to-find virus or bacterium. Even today, some textbooks still hedge and hint at the possibility that autoimmune hepatitis could be due to an undetectable pathogen. But in fact, scientists have now isolated the provoking antigen in autoimmune hepatitis; it is called "liver-specific protein," or LSP, and is a perfectly normal part of healthy liver cells.

Although autoimmune hepatitis has no proximal connection to viral infection, it cannot be ruled out that a previous viral infection, perhaps even with one of the hepatitis viruses, selected a coterie of liver-specific memory T cells that subsequently became involved in autoimmune damage. And although numerous antibodies to self liver proteins are present in autoimmune hepatitis, there is no evidence that these contribute to liver damage. Persons with certain MHC types are more likely to develop autoimmune hepatitis, a sure sign of T-cell involvement, and the most effective therapies target T cells.

SYSTEMIC LUPUS ERYTHEMATOSUS

SLE, or "lupus," is the classic example of an autoimmune disease in which the immune system attacks not a specific tissue or organ in the body, but rather a wide range of self tissues. Like most autoimmune diseases of this type, lupus is seen most frequently in females, almost always setting in during the peak reproductive years. Lupus is 10 times more common in women than men, and more common in people of color than in Caucasians. The *erythematosus* in SLE refers to a rash that often breaks out on the face, particularly around the nose. This so-called "butterfly rash" is one manifestation of the general sensitivity of lupus patients to ultraviolet light, including sunlight. Other symptoms include fever, weakness, pleurisy, anemia, and heart and kidney problems. Joint pain from arthritis is a common concomitant of lupus throughout all its stages. No one really knows what causes it, but it is often associated with recurrent Epstein-Barr virus infections and certain prescription medications.

Lupus is accompanied by antibodies to a wide range of self antigens, one of the most unusual being DNA. Although many other autoantibodies (e.g., against thyroid or liver tissue; muscle; and blood cells and serum proteins) are found in lupus patients, antibodies to DNA are the most prominent, and are in fact diagnostic for the disease. It is likely that the DNA antibodies are formed against DNA released by dying cells. It is not clear whether the DNA antibodies themselves cause any harm. Such antibodies are not formed in other diseases in which cells die and release their contents, so their appearance in lupus is still something of a mystery. If we knew why these particular antibodies were formed in the first place, we would likely understand a great deal more than we do about this disease.

Like other antibody-mediated autoimmune disorders, much of the serious damage in SLE comes from the deposition of antigen–antibody complexes, not consumed by macrophages, into blood vessels throughout the body. When this occurs in blood vessels in the kidneys, for example, a condition known as **glomerulonephritis** can develop, which eventually may lead to serious kidney problems and even kidney failure. Because the antigens in lupus (and other autoimmune diseases) are a part of self, there is in effect an endless supply of them, and an endless stream of immune complexes just keep on forming. In advanced cases, lupus may also affect the nervous system. This can result in pain throughout the body, but may also result in actual damage to the central nervous system, manifesting as headache, paralysis, seizures, or other neuropsychiatric problems.

Lupus is not really curable. It can be controlled in many cases with steroids such as prednisone. Mild immunosuppressive drug treatments can also help. But these kinds of drugs are not without their own risks. Arthritis and kidney problems often worsen with age, causing considerable distress and affecting the general quality of life. On the other hand, with careful monitoring by an experienced physician, it is not obvious that lifespan per se is greatly affected by diseases such as lupus.

MYASTHENIA GRAVIS

Myasthenia gravis (MG) is a disease characterized by extreme muscular weakness, usually beginning in the head and neck but in most cases extending to the entire body. It is twice as frequent in women as in men, and is seen earlier in women (average age of onset 28 years, vs. 42 years in men). The disease in men is often more limited as well. The first visible signs of myasthenia are usually drooping eyelids and sagging neck and facial muscles. Patients may experience difficulty in breathing and swallowing, and may have vision problems as well.

Myasthenia was recognized as far back as the midseventeenth century as a distinct condition, although its autoimmune basis could not of course have been known. The following description, written by the English physician Thomas Willis in 1672 in his *De Anima Brutorum*, pointed to an affliction that often accompanies the onset of this disease:

. . . she for some time can speak freely and readily enough, but after she has spoke long, or hastily, or eagerly, she is not able to speak a word, but becomes mute as a fish, nor can she recover the use of her voice under an hour or two.

It was only in 1934 that drugs that relieve the most severely debilitating symptoms of MG were discovered. With the development of artificial respirators a few years later, the world saw a rapid drop in mortality from this disease by 1940. Before that time patients went largely untreated and often died from respiratory failure within a year or so of onset. Currently, MG is fatal in only about 10% of those afflicted, although it is never curable.

The defect in MG is an interesting one and involves one of the most highly restricted antiself attacks of any of the autoimmune diseases. Patients with MG make antibodies that affect the response to a neurotransmitter called acetylcholine (ACh). ACh is released from the tip of a nerve cell at the point where it attaches to a muscle and is picked up by a special acetylcholine receptor (AChR) on the muscle being served. This causes the muscle to contract and carry out its function.

MG patients make antibodies to their own AChR; these antibodies block the muscle's ability to pick up and respond to ACh. There is nothing wrong with the muscle per se; it simply cannot be stimulated by the nervous system to do its job. In animal models of this disease, passing the antibody from an animal with MG to a healthy animal is sufficient to pass the symptoms of MG. Pregnant women may pass the antibodies to their developing child, which may be born with symptoms of the disease (the symptoms fade within the first few months of life). So in this instance a single antibody, specific for a single target molecule (AChR), appears sufficient to explain an entire disease.

Yet, in spite of the narrowness of the immune attack in MG, most patients do show signs of a more generalized autoimmunity. As many as a third will have clinically detectable Graves' disease, which affects the thyroid. There is little to suggest Graves' disease is caused by the same antibodies that cause MG; if it were, then *all* MG patients should have Graves' disease.

DIABETES

Type 1 insulin-dependent diabetes mellitus (IDDM) is a chronic autoimmune disease in which the immune system gradually destroys the insulin-producing β-cells in the pancreas. The primary result is loss of the ability of cells in the body to take up sugar, a primary nutrient. Prior to the development of injectable insulin, this disease was almost uniformly fatal.

IDDM has a strong genetic component. However, concordance for IDDM in identical twins rarely exceeds 50%, suggesting the involvement of one or more environmental "triggers" in onset of disease. One such environmental factor could well be viral infection. Serotype B Coxsackie virus in particular has been implicated in triggering IDDM in susceptible individuals. Very often newly diagnosed patients will have recently experienced a Coxsackie virus infection or display Coxsackie antibodies in their serum. To the extent that viruses play a role in onset of IDDM, we may expect similarities in the immunopathologies of IDDM and certain chronic viral diseases.

T cells are the major destructive agent in IDDM. The antigens in the pancreas against which autoimmunity is directed are probably diverse, but at present the major antigen identified in both mice and humans in eliciting cell-mediated responses is **glutamic acid decarboxylase (GAD)**. While GAD is not islet specific, the damage produced during diabetes appears relatively islet restricted. Experiments in mice showed that the severity of the disease correlated directly with the level of GAD expression. Intriguingly, Coxsackie virus infection causes increased GAD expression in the pancreas; moreover, one of the Coxsackie-encoded proteins contains a peptide region with strong homology to a region of the GAD protein.

Insulin itself also appears to be a target of T cells in human IDDM. CD4 cells that recognize human insulin can be isolated from abdominal lymph nodes in diabetics and shown to produce inflammatory cytokines when presented with insulin in vitro. Whether these T cells were present before diabetes developed, and were responsible for the onset of the disease, is still unclear. They may have arisen as the pancreatic β-cells began to collapse from other causes.

A wide range of immune components are mobilized in IDDM, but the damage leading to loss of insulin production is attributable mostly to CD4 and CD8 T cells. Early in the disease, macrophages and dendritic cells infiltrate the pancreas, followed by T cells, B cells, and NK cells. The T cells cluster around and physically penetrate the pancreatic islets. Eventually, β cells within the islets are selectively killed, and insulin production is compromised. Islet cell death is apoptotic.

Both T cell subsets are able to passively transfer at least some aspect of the disease. While immune CD4 cells, like immune CD8 cells, can accelerate development of disease, only immune CD8 cells are able to actually kill islet cells in vitro.

Interestingly, there is no gender bias in IDDM; diabetic patients are split just about equally between men and women. This is unusual for a human autoimmune disease. At present, the implications of this observation for the development of IDDM in individual patients are unclear.

RHEUMATOID ARTHRITIS AND THE MIND-IMMUNE SYSTEM CONNECTION

There is a great deal of data showing that the mind—the brain—and the immune system communicate on a constant basis, exchanging information via cytokines and neurotransmitters. One of the earliest indications that this might be so came from a remarkable series of studies in the 1960s on an autoimmune disease, rheumatoid arthritis (RA).

RA is definitely autoimmune in nature. Patients with RA make a type of antibody called *rheumatoid factor* that is not specific for an organ or a tissue, but for other antibodies! Aside from the problems this could cause for antibody function, it also leads to the formation of truly large amounts of immune complexes. As the "aggressor" antibodies (rheumatoid factor) collide with and bind to innocent bystander antibodies in the blood, large complexes consisting of antibodies recognizing and binding to each other are formed.

Normally such complexes are efficiently cleared away by macrophages. But as in lupus, when the amounts of immune complex exceed the ability of macrophages to clear them from the bloodstream, these complexes can be deposited on the inside lining of blood vessels or, in the case of RA, in the joints. An inflammatory reaction follows. As T cells, B cells, and macrophages enter the joints and try to clear the antigen–antibody complexes away, the smooth tissue that helps lubricate the interaction of bones within the joints is gradually destroyed. This process is painful and, over time, deforming to the joints—the disease we know as arthritis.

Like many other autoimmune diseases, RA has a marked genetic component; it tends to run in families. The ratio of female to male patients is very high. But there had been persistent reports in the RA literature that there might also be an emotional or "personality" component as well. Patients with RA were consistently described (by their doctors, their family members, and themselves) as "tense," "moody," and "high-strung." They tended to have very strict standards for themselves and others, and reacted negatively when they perceived that those standards were violated. The problem was that the data on psychological contributions to RA were difficult to interpret. They had been collected by researchers in a wide range of disciplines—internal medicine, psychiatry, psychology —each with their own technical approach and particular point of view. Still, a common, underlying theme persisted.

In the face of these intriguing but largely unsubstantiated elements of "common wisdom," Drs. George Solomon and Rudolph Moos of Stanford University's Department of Psychiatry carried out a detailed and carefully controlled analysis of groups of female RA patients. They were intrigued by, among other things, a recent comparison of genetically identical female twins, only one of whom in each instance had clinically diagnosed RA. Clearly in such cases, both twins had identical genetic constitutions; why then did only one sister develop RA? In this particular study, it was found that the twin who developed the disease had had a recent, serious interpersonal conflict accompanied by considerable psychological stress. The authors of the study suggested that development of RA might actually have been caused by an interplay of both genetic and emotional factors.

For their own study, Solomon and Moos chose to analyze not only women affected by RA, but also the nearest-aged healthy female siblings of the RA patient as controls. Applying a wide range of written tests, oral interviews, and clinical examinations, they produced a convincing set of insights into the relation between emotional states and susceptibility to RA. Their data supported some of the previously held notions about this disease, while refuting others. They did not find, as others had previously suggested, that women with RA were more physically active, concerned about their appearance, or dependent in relationships.

They did find, however, that in nearly all cases the sisters with RA tended to be more nervous, more depressed, or quicker to anger in reaction to a real or imagined slight than their symptom-free siblings. In almost every case, emotional conflict correlated either with the onset or with a pronounced worsening of the disease. Close questioning of the patients and their family members suggested that these traits were not brought on by the burden of the disease itself, but were personality characteristics of the patients before the disease set in.

In a subsequent study Solomon and Moos took a closer look at the healthy sisters of their RA patients. A number of them showed evidence of rheumatoid factor in their blood, suggesting that they had inherited the same genetic predisposition to RA as their affected sisters. In some cases these levels were even within the range found in patients with active RA. Why then had these women not developed the disease? Psychological testing showed them to be almost exactly opposite in personality type to their siblings with RA. They were generally happy, outgoing individuals who either managed to avoid potentially stressful situations or who coped well with them once they developed.

Solomon and Moos concluded from their studies that in some fashion the mind, as manifested in personality, is able to exert a modulating influence on the immune system that can either favor or discourage the initiation or progression of an autoimmune disease, rheumatoid arthritis. This is now a generally accepted notion about the development of autoimmune diseases such as RA, lupus, and multiple sclerosis, among others: they may not always represent a failure of the immune system per se, but may reflect a combination of an immune abnormality exacerbated by emotional stress.

These studies suggest that the mind can exert a direct influence on the immune system itself, in this case helping determine whether or not an autoimmune disease developed. This may be akin to the influence the mind apparently exerts on other specific organ systems—for example, the increase in cardiovascular problems seen in persons mourning the loss of someone very close. But the immune system is unique among organ systems of the body in that it is instrumental in maintaining health. Is it possible that the mind exerts an even greater influence on human health by acting through the immune system?

EVEN THE COMMON COLD...

The most convincing demonstration that the mind—in its perception of and response to stress—can directly influence the body's immune response to a foreign pathogen comes from an interesting study measuring responses to the common cold. A group of researchers prospectively analyzed 394 physically healthy volunteers to determine their current psychological stress status before deliberately exposing them to a series of cold viruses. Some of the parameters used to evaluate stress levels included recent loss of a close friend or relative; the degree to which an individual felt that current demands in his or her life exceeded the ability to cope; and (as in the RA study) the extent to which a subject described himself or herself with words such as "nervous," "angry," "depressed," "dissatisfied," etc.

Using a composite of all these parameters, the volunteers were grouped in categories ranging from very low to very high stress. Great care was taken to be sure that these categorizations represented the subjects' stress levels at the time of the test and were not generalizations about the subjects' responses to stress at other times in their lives.

After completion of psychological evaluation and after being fully informed of the risks they were about to be exposed to, the volunteers were given nose drops containing a low infectious dose of one of five different common cold viruses. They were then housed in special apartments and monitored daily by a physician. Small samples of nose tissue were collected by swabbing to determine whether or not the virus had succeeded in establishing itself, and each subject was observed closely for standard cold symptoms.

The rate at which subjects became infected with the viruses, and the rate at which they developed clinically verifiable colds, correlated exactly with their stress levels. For example, 27% of the individuals judged to have little or no stress developed colds; nearly 50% of those in the high-stress category developed clinical cold symptoms. The rate at which infection and colds developed had absolutely no correlation with a wide range of other parameters such as age, sex, education level, smoking habits, alcohol use, exercise, or sleep habits. This study left little doubt that negative psychological states, and the stress they engender, can weaken the body's resistance to infectious disease, as well as exacerbate internal problems such as autoimmunity.

Studies like these, which have been confirmed many times over, gave rise to a new branch of immunology—**psychoneuroimmunology**—with its own meetings and scientific journals. It is now commonly accepted that the brain regards the immune system in the same way it regards sight, sound, smell, and all the other senses—as a source of information about what is happening, in this case, mostly inside the body. In fact, the immune system is sometimes referred to as the mind's "sixth sense."

Not only does the brain gather information about things like infection and try to help by speeding up production of white cells or raising temperature to inhibit bacterial growth, but it can also directly modulate and interfere with immune responses themselves—and not always in a helpful way, as we have seen in the case of RA. We all know of instances where someone who recently lost a spouse or a child became seriously depressed, perhaps came down with a serious medical condition, and possibly passed away themselves. Clearly depression affects many of the body's physiological systems, and the immune system is no exception. Direct measurements have shown that both T-cell and B-cell function are significantly inhibited during depression. Why this should be so, what it is intended to help, is not at all obvious.

WHY AUTOIMMUNITY?

Who needs autoimmunity? Where does it come from? It could be viewed as just another way nature has of being sure we don't hang around too long, using up valuable resources. But in fact, with a few exceptions most autoimmune diseases are not all that life threatening. They make life miserable, but they don't usually kill us. So how do they fit into the grand scheme of things? Why does the immune system turn against self?

Although many autoimmune diseases seem almost certainly to

represent an unprovoked attack of the immune system on self, the possibility that at least some such diseases are due to cryptic microorganisms continues to intrigue many immunologists. If even tiny traces of invading microbes remain lodged in human tissues after an infection, they argue, immune-based disease could ensue. Although the microorganisms would be present in amounts too low to be detected by even the most sensitive clinical tests, they would still be detected by the immune system. In such small amounts, even the most virulent microbes would themselves be unlikely to cause disease, but the attempts of the immune system to ferret them out and destroy them could cause extensive damage to apparently normal human tissues. The problem with such hypotheses, of course, is that they are virtually impossible to either prove or disprove, since they posit things that cannot be measured. But that doesn't mean we shouldn't keep looking!

An interesting variant of this hypothesis is something called **antigenic mimicry**. What if an invading bacterium or virus contained a protein, a very small region of which was identical to some human protein? In the process of responding immunologically to that particular stretch of the foreign bacterial or viral protein, might we produce antibodies or activated T cells capable of attacking the corresponding human protein? This is very likely to be the case in Type I diabetes. **Rheumatic fever** (rheumatic carditis) is another autoimmune condition in which we produce antibodies against our own heart proteins. This disease almost always follows on the heels of a previous infection with streptococcal bacteria. Although the antibodies causing the damage are clearly directed against human heart muscle proteins, it had been suspected for years that the antigen triggering the antibodies was actually streptococcal in origin. Scientists have now isolated a 32 amino acid segment of one of the surface proteins of streptococcal bacteria that induces the antibodies that cross react with human heart muscle.

So quite likely some diseases that we think of as autoimmune may be various forms of spillover from normal immune attacks against foreign invaders. But equally likely, we may just have to come to grips with the possibility that the immune system does, on occasion, decide to attack self, unprovoked by outside agents. Is this simply one more cross we must bear, one more price we must pay for an immune system that does a pretty good job most of the time? Or could it be that autoimmunity is a normal part of human biology, playing a more profound role than malicious aggravation?

A close pursuit of this very question has led to some intriguing insights into how the immune system is put together. For example, it has been observed that the immune system, both in terms of T cells and of B cells, seems to be directed, right around the time of birth, largely against self. If we examine the antibodies in the blood of human infants just after birth, we find that a rather high percentage of them are directed at self antigens. This condition disappears a short time after birth, but it is as if, just prior to that instant when the immune system was taking that last look around to define "self" at birth, it was actually using self antigens to prime itself, to get itself up and going.

This phenomenon is thus probably connected to the issues of tolerance and fetal development discussed earlier in this chapter; the immune system is busy investigating what is and is not self. As far as we can tell, this self-reactivity causes no harm, either in the fetus or in the newborn. But beyond being simply a neutral phenomenon, this observation has prodded scientists to wonder whether in fact this mild form of self-reactivity by the immune system may actually be a necessary and beneficial step in the development of the fetus. So both at the very beginning and the very end of life, we see significant levels of self-reactivity by the immune system. Right now, nobody knows what that means, but you can be sure it is a question that will continue to be pursued.

APPROACHES TO TREATING AUTOIMMUNE DISEASE

Current therapies for autoimmune disease are not terribly effective. For the most part they are based on mild immunosuppression, aimed at the B-cell or T-cell arm of immunity, depending on the disease. Corticosteroids, cyclophosphamide, and cyclosporin are among the drugs used. But these are not disease-specific drugs by any means, and they depend on a generalized rather than specific suppression of the immune system. As such, they risk suppression of responsiveness to microbial antigens as well as reactivity to self, and can lead to emergence of opportunistic infections as well. The goal of these drugs is not to cure the underlying disease, but simply to manage it so that life becomes a bit more tolerable for afflicted individuals.

A new approach for not just managing, but curing, autoimmune disease is based on hematopoietic **stem cell autotransplantation**. This approach, which has worked well in animals and is now in human clinical trials, is based on the following knowledge and assumptions. First, we know that what a T cell or B cell recognizes as foreign depends on what kinds of receptors T cells and B cells randomly generate. Some of us will randomly generate receptors that cross react with self molecules, but, because this is a completely random process (chapter 2), each of us—even genetically identical twins—will generate different subsets of self-reactive cells. Each of us also generates different groups of receptors that are reactive with various environmental antigens, such as microbial antigens.

Autoimmunity arises, we think, because the tolerance mechanisms acting to control self-reactive T and B cells that escape elimination in the bone marrow and thymus break down, allowing these cells to begin reacting against self. The interaction of such cells both with regulatory mechanisms and with self antigens is likely affected by the particular fine specificities of the receptors involved. Alternatively, T and B cells in a particular individual that are selected because they responded to a particular environmental antigen may happen to cross react with a self-antigenic epitope (antigenic mimicry). As memory cells are built up to the environmental antigen, some of those that are cross reactive with self may become difficult or impossible to control. Again, this likely reflects at least in part the particular receptor fine specificities the individual has generated.

If the extent to which either of these causative mechanisms becomes a problem in a particular individual is influenced by the particular array of randomly generated receptors he or she happens to produce at different stages in life, then why not just erase that particular array and start over?

The idea is fairly simple. Under protected conditions, remove samples of an individual's bone marrow and enrich for the hematopoietic stem cells that can replenish the entire immune system. Then, using radiation or drugs or some combination of the two, erase most of the existing T and B cells in that individual, particularly memory cells, as well as much of the bone marrow. Then, reinfuse the hematopoietic stem cells back into the individual and let the adaptive immune system regenerate itself, coming up with a different collection of T- and B-cell receptors. Let these new T and B cells interact with the regulatory mechanisms suppressing self-immunity and build up new sets of memory T and B cells.

It sounds simple, and almost too good to be true. But in fact, it has worked very well in animals that have spontaneous forms of autoimmune disease. It has also been possible to use stem cells from genetically different individuals to reconstitute the immune system **(allogeneic stem cell transplantation)**, provided that the immune system of the recipient is thoroughly suppressed before the foreign stem cells are infused. Clinical trials involving several variants of this approach are now in progress, and we are awaiting their evaluation. Should this work as well in humans as it has in animals, we may, for the very first time, have a cure for many debilitating autoimmune diseases, rather than temporary palliation.

13

Organ Transplantation

Late in the summer of 1954, Richard Herrick was referred by his doctor to the Peter Bent Brigham Hospital in Boston, Massachusetts. Richard was 24 years old and had been suffering for some time from high blood pressure and puffiness around the face and eyes. His doctor suspected a kidney problem, and the Brigham is where you went if you were worried about your kidneys. The medical staff at the Brigham ran a battery of tests that initially might have indicated any number of problems. But they noticed that in addition to high blood pressure, Richard had more protein than normal in his urine, as well as traces of blood. Together with other findings, this confirmed the diagnosis of a kidney dysfunction. Richard was transfused with several units of blood, which improved his condition considerably, and he was sent home. Only time would tell how serious the problem with his kidneys was.

Five months later, Richard Herrick was back, and this time it was clear he was in trouble. His blood pressure was dangerously high. Protein levels in his urine were double what they had been before, and he was showing signs of congestive heart failure. Several days after this second admission, Richard began to exhibit bizarre behavioral changes; he occasionally became drowsy and disoriented; at other times he was irritable or even aggressive toward the staff. He went into convulsions several times. It was a set of symptoms the doctors at the Brigham were all too familiar with, and about which they knew they could do precious little. Their young patient was experiencing the beginning stages of massive and terminal kidney failure.

Dr. John P. Merrill took a special interest in this particular patient. Merrill had been working with a medical equipment company on the refinement of an "artificial kidney," what we would today call a renal dialysis machine. This machine, first developed in Holland during World War II, was showing great promise in being able to substitute for one of the most vital kidney functions—removing from the blood toxic substances that could cause precisely the symptoms this young man was experiencing. In fact, on his second visit to the hospital Richard was treated with one of the artificial kidneys and, as the doctors expected, showed great improvement.

But another chance to demonstrate the usefulness of his new machines was not what attracted Merrill to this case. Merrill knew that the kidney machine could never be more than a stopgap measure, able to keep a patient alive for a period of time but never able to offer a cure. What he was really interested in was the possibility of kidney transplantation. He had recently completed a series of nine kidney transplants, taking healthy kidneys immediately after death from patients who died of causes unrelated to their kidneys and transplanting them into patients with terminal kidney failure. In several cases, the transplanted kidney had seemed to take hold for a while, bringing almost immediate improvement in the recipient's condition. But in a fairly short time all nine transplants had failed, and the recipients all ultimately died of kidney failure.

Like other experts in his field, Merrill was convinced the transplants were failing not because of problems with the surgery or because an organ from one person simply could not function in another, but because the transplanted organ was being attacked and rejected by the recipient's immune system. Merrill had argued for some time that human identical twins should be able to exchange organs and tissues without any fear of immunological rejection. Inbred mouse strains, which are like human identical twins, could do it.

And that was what interested Merrill about this young man. According to the doctor who had referred Richard Herrick to the

Brigham for treatment, Richard had an identical twin, Ronald. After reassurances that he could survive with a single kidney, Ronald agreed to give the new procedure a try. Dr. Merrill and the Herrick boys were about to make medical history.

As a preliminary test of his hypothesis, Merrill's team carried out an exchange of skin grafts between Richard and his twin brother. After a rather anxious month in which his doctors had to struggle to keep Richard alive, it was confirmed by microscopic examination that he had completely accepted Ronald's skin. Without waiting any further, the two brothers were prepped and wheeled into adjacent operating rooms. Ronald's left kidney was removed and taken in a stainless steel pan to the surgeons waiting in the adjoining operating room. While the first twin was being closed, the surgeons opening Richard saw a sight usually only seen at autopsy—two shriveled, shrunken kidneys, wasted away to a tenth their normal size.

Although Ronald's kidney had grown pale and cold during the 80-odd minutes between operations, as soon as it was connected to Richard's blood system it swelled ever so slightly and turned pink and warm to the touch. After the surgeons checked meticulously for leakage, this young man, who only days before had been within a stone's throw of death, was carefully sewn back together. Recovery from the surgery was uneventful for both brothers, and the transplanted kidney began to function beautifully in its new surroundings.

All of Richard's previous symptoms disappeared in a matter of days. He was discharged after two weeks, and over the course of the next few months regained his former physical vigor, as well as 25 pounds of lost weight. Ronald's remaining kidney underwent a gradual enlargement as it took on the sole task of cleaning out his blood, but he suffered no ill effects whatsoever. Both brothers lived for many years.

And so began the age of human organ transplantation. Of all the miracles wrought by modern medicine, none has moved us quite the way organ transplantation has. That an organ can be severed of all its connections with one human being, implanted into another, and recover the full function it needs to sustain life in the recipient was and remains simply awe inspiring. In the case of bone marrow or a kidney, or a lobe of liver, both the donor and the recipient may be alive and well after the transplant has been accomplished. A bond is established between them that is unique in human experience. On the other hand, to see a transplanted heart still beating and sustaining life in a human being a quarter century after its original owner has died puts us in close touch with some of the deepest mysteries of life and stretches our conception of the meaning of mortality and immortality. How did we come to be able to do such a miraculous thing?

THE IMMUNOLOGY OF ORGAN TRANSPLANTATION

As we begin our consideration of the role of the immune system in the rejection of **allografts**—organs, tissues, or cells exchanged between members of the same species (Table 13.1)—it is good to bear in mind that here we are talking not about an immune function that occurs in nature, but a situation the immune system was never presented with in its entire evolutionary history. And the intent of our manipulation of the immune system in the case of transplantation is different than it was in other situations we have examined, except autoimmunity. Instead of trying to get the immune system to work harder, or more accurately, our intent is to stop it from doing what it thinks it should be doing. The challenge for us is how to get it to stop doing its job in the case of life-saving transplanted body parts without knocking out its ability to do what nature intended it to do—protect us from a world of microbial predators.

In rejecting a transplant, the immune system basically has to make it up as it goes along. As we will see, the immune system—and in the case of allograft rejection, we are talking mostly about the T-cell branch of the adaptive immune system and possibly its

TABLE 13.1
Types of Grafts That May Be Exchanged Between Individuals

Autograft	A graft taken from one part of an individual and transplanted to another part of the same individual
Isograft	A graft exchanged between two genetically identical individuals (identical twins, two members [of the same sex] of inbred animal strain)
Allograft	A graft exchanged between two nonidentical members of the same species
Xenograft	A graft exchanged between members of two different species

ability to drive inflammation—will do what it does in most cases, and that is blindly attack anything it perceives as not self or altered self. It can be forgiven for not knowing that an incoming kidney could save your life.

The rejection of an allografted organ was yet another example of those mysterious reactions that seemed immunological in nature but couldn't be tied in to what was known about the immune system in the first half of the twentieth century. Sir Peter Medawar, in work he did on burn victims during World War II in London, observed that if a person or animal received a skin graft twice from the same source, the graft would be rejected much more rapidly the second time around. This was a powerful argument supporting the involvement of the immune system. (And this is the last time we will mention it—his work eventually led to a Nobel Prize in 1960.)

Medawar and others followed up these observations in the laboratory in the late 1940s and early 1950s. If a mouse from inbred strain A is given a skin graft from a mouse of the genetically distinct strain B, it will be rejected in 11 to 13 days. If the A mouse is given another B graft a few weeks later, it will reject it in five to six days. But if you give this same A mouse a skin graft from strain C, it will reject it in 11 to 13 days. "Looks like immunological memory to me," everyone said, and indeed there were anti-B antibodies in the serum of the A mouse that had just rejected a B graft. But when those antibodies were transferred to a naïve A strain mouse, which was then grafted with B skin, rejection still took 11 to 13 days. The antibodies were doing nothing. It wasn't long before someone figured out that you can passively transfer immunity to allografts with cells, but not antibodies. And of course, the cells turned out to be the CD4 and CD8 T cells you're already familiar with.

THE PROBLEM OF HISTOCOMPATIBILITY

It might seem intuitively obvious that human beings are all very different from each other and that the immune system would naturally spot these differences and respond to them. But what exactly are the differences between people that the immune system responds to? These differences are clearly absent in identical twins and present in everyone else. But are all differences the same? Might some people be closer in terms of these differences than others? And if so, is it easier to exchange grafts between them?

Figuring out exactly what it is that is recognized when one person rejects another's tissues or organs was actually a spin-off of the early immunology of cancer studies we talked about in chapter 11. Once researchers realized that rejection of tumors between two people was just an example of transplant rejection, they used tumor allografts as a model system for studying the immunology of transplantation in animals.

During the course of these studies, they noticed that the more closely related two mice were, the slower was the rejection rate for transplanted tumors. That led to the development of inbred strains through brother–sister matings. The more generations of inbreeding, the closer the offspring became genetically and the slower the rate of rejection when they exchanged grafts, until after 20 generations or so they could exchange cells, tissues, and organs with impunity. Members of inbred strains had become, essentially, genetically identical twins.

The same is true for humans. The more closely related two individuals are, the less vigorous is the rejection reaction when they exchange body parts. Sometimes closely matched siblings can exchange kidneys or bone marrow with only a mild, easily controlled rejection response. But in some instances a sibling donor's organ may also scarcely do better than one selected from an unrelated donor.

Using their inbred strains of mice to probe the genes that, when different between donor and recipient, most influenced rejection, researchers gradually zeroed in on the problem. At first they thought there would be only one gene, or a few at most. They were *very* wrong! It took nearly half a century to sort it all out, and it was far more complex than anyone could have imagined. The work of literally thousands of researchers led finally to the discovery and full characterization of **histocompatibility antigens,** and the genes that encode them, which are found grouped together in the **major histocompatibility complex,** or **MHC** (Figure 13.1A). The MHC genes in each species have a different name—in humans we refer to MHC genes and proteins as **HLA** genes and proteins.

It was obvious from the beginning that MHC genes and products cannot have as their primary function triggering of graft rejection, since exchange of tissues and organs never occurs in nature. Why would we have a portion of our immune systems dedicated to something that only happened in the last 60 years of our evolutionary history? So when Rolf Zinkernagel and Peter Doherty showed that the real job of MHC proteins is to present peptide antigens to T cells (chapter 4), there was a huge collective sigh of relief among both research immunologists and transplant surgeons.

And that idea made sense. T cells are selected in the thymus to recognize and interact with self MHC proteins, either class I or class

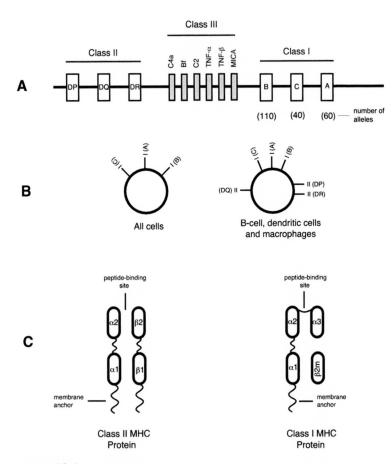

FIGURE 13.1

The histocompatibility system in humans, HLA. **A.** HLA genes are aligned along the short arm of chromosome 6. **B.** All cells (except red blood cells) display class I MHC proteins. Class II proteins are found only on dendritic cells, B cells, and macrophages. **C.** Class I and class II HLA proteins are involved in antigen presentation. Peptide binding on class II proteins occurs in the space between the α-2 and β-2 domains of the two adjacent chains, which are not linked together. Class I molecules bind antigen at a site between the α-2 and α-3 domains of the class I chain.

II, in order to be able to look at peptides bound to them. When a transplant comes into your body, almost immediately your T cells begin nosing around the incoming MHC molecules on the incoming cells. The MHC proteins on the transplant are *sort* of like your own, but not really. And the peptides they display are not from you, either!

That's enough for T cells! From their point of view, something is clearly not quite right, and T cells are not ones for taking chances. If it's not you, it's got to go—"my way or the highway," so to speak. So your T cells immediately set about destroying the transplant, in the same way they would attack and destroy one of your own cells or tissues that had become virally infected or cancerous.

MHC GENES AND PROTEINS

There are three distinct groups of genes in the human HLA gene complex (Figure 13.1). The class I and II MHC genes you already know about. In addition, there is a large group of genes loosely referred to as class III MHC genes, interspersed among the clusters of class I and II genes. Nearly all of these genes play some role in the immune system, but they have quite different functions. Some, like TAP-1 and LMP2, are involved in loading peptides onto class I proteins. C4a and Bf are complement components. Tumor necrosis factor-α (TNF-α) and TNF-β are cytokines. None of the class III proteins are found at the surface of cells, like the class I and II proteins.

Humans have three different class I MHC (HLA) proteins, all of which are present at the same time on the surface of all cells in the body (Figure 13.1B). These are called HLA-A, HLA-B, and HLA-C. The genes for these three proteins are lined up on human chromosome six as shown in Figure 13.1A. Since humans have two copies of every chromosome, one inherited from the father and one from the mother, we each have a total of six class I molecules. Each of these can be present in hundreds of copies on any given cell.

Each class I protein consists of a single α chain composed of three domains, α1, α2, and α3. The peptide-binding site is located at the top surfaces of the α2 and α3 domains, which are physically connected. The α chain is associated at the membrane with a molecule called β2 microglobulin, which basically acts as a fourth domain to keep the class I molecule in the correct shape, but plays no role in peptide binding (Figure 13.1C). The class I molecule is anchored into cell-surface membranes by a peptide tail hanging down from the α1 domain.

Like nearly all MHC genes and proteins in all species, the human HLA class I genes and proteins are incredibly **polymorphic**. That means that numerous mutations have been allowed to creep into the genes, so that there may be a hundred or more different forms **(alleles)** of each of the class I genes scattered throughout the species. The number of alleles of each gene, combined with the fact that we have two separately inherited sets of these genes, means that the likelihood of finding two randomly selected people with the same class I proteins is about 1 in 70 billion. Since there are only 6 billion people in the world, it is virtually impossible to get a complete class I match with a random unrelated donor. Brothers and sisters, on the other hand, can on occasion have the same class I gene alleles, and thus the same class I proteins.[1]

The class II MHC proteins are even more complex. As shown in Figure 13.1A, there are three major gene groups encoding class II proteins in humans (HLA-DP, HLA-DQ, and HLA-DR). The class II HLA proteins are also all present at the same time, but only on a limited number of cells that interact with T cells: B cells, macrophages, and dendritic cells. These latter of course also have the three class I HLA proteins.

---

1. But even if they do have exactly the same display of parentally inherited class I and II proteins at their cell surfaces, they will be different at many other genetic loci and will be displaying different peptides on their otherwise identical HLA molecules.

Each class II protein consists of two separate chains, α and β each encoded by a separate gene. The peptide-binding site is formed by the α2 and β2 domains, which are adjacent though not physically connected. Each of the component chains is anchored separately to the cell-surface membrane. Each of these class II genes also has dozens if not hundreds of alleles in humans. The bottom line is that the odds of matching two randomly selected individuals for both class I and class II HLA genes—getting a perfect match for a transplant—is about as close to zero as you can get. Family members have a better chance of being matched at least partially, and identical twins are matched perfectly.

So MHC proteins truly are markers of individuality within humans, as within all vertebrate species. But the polymorphism of MHC genes and proteins has important consequences outside of organ transplantation. The true function of MHC proteins of course is in the presentation of fragments of peptide molecules to T cells. This involves chemical interactions between the antigen-binding sites of MHC proteins and individual peptide fragments. The fact that each of us has a different combination of class I and class II MHC molecules means that each of us will react slightly differently with the antigenic universe around us in terms of presentation of peptides to our T cells. As we grind up the proteins of invading pathogens, each of us will select slightly different combinations of these peptides for inspection by our T cells. Thus, each of our immune responses will be slightly different; some of us will respond more effectively to some pathogens and less effectively to others. As a species, we'll be pretty well covered, but some of us may have holes in our defenses.

HISTOCOMPATIBILITY TESTING IS MANDATORY FOR TRANSPLANTATION

There is not always a well-matched family member available for organ donation, and apart from kidneys, bone marrow, liver, and a few other tissues, living donors are out of the question anyway. (Livers can come from a living donor, because only a portion of the liver is enough. The removed portion grows back in the donor, and the donated portion expands into a full liver in the recipient.)

Most organ donations come from recently deceased individuals who have indicated a willingness to have their organs used for transplantation into others. On rare occasions, an organ may come from an unrelated living donor. Whoever the donor is, it is absolutely essential to determine the degree of tissue compatibility between donor and recipient. This is done by a process called *histocompatibility typing*, in which the different possible varieties of MHC antigens of prospective donors and recipients are identified and the best possible match is made. There is a good correlation between the degree of MHC matching and the success of the transplant. Especially when the donor and recipient are unrelated, every effort is made to achieve the closest possible HLA match between them.

Still, even with the best-matched transplant (unless you're lucky enough to have an identical twin, and don't need a heart), in most cases rejection will eventually occur. Mild rejection responses can be managed fairly well with immunosuppressive drugs like cyclosporin A or rapamycin, and most transplant recipients now routinely live a decade or more with their transplants (Figure 13.2). But that is not without a price. All of the drugs used to inhibit rejection do so by suppressing some aspect of T-cell function, and that leaves us open to both external and internal (opportunistic) infections.

The opportunistic pathogens that make their appearance in immunosuppressed transplant patients are similar to those seen in AIDS and severe combined immune-deficiency disease (SCID) patients. The only way to manage this is to temporarily decrease the immunosuppressive drugs while the T cells recover and chase the opportunistic pathogens back into hiding. So long-term transplant patients are always engaged in a dance of death with their pathogens and their transplants, and occasionally run into serious trouble.

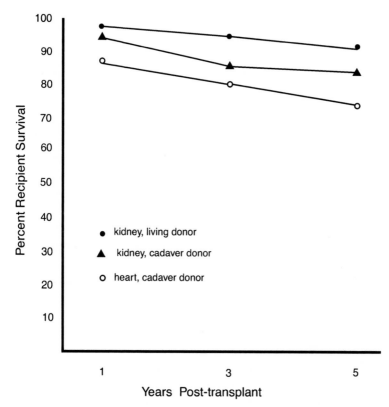

FIGURE 13.2
Recipient survival in the years immediately following transplantation.

Some immunosuppressive drugs can also predispose toward cancer, although this is less of a problem with some of the more recent drugs. But transplant physicians have now accumulated five decades of experience dealing with these problems, and the number of lethal failures, especially with reasonably well-matched recipients, is considerably less than in the beginning.

On the other hand, it must be admitted that the rate of increase in transplant survival has tapered off in the past decade. There are various reasons for this. Although better immunosuppressive drugs are available, waiting lists for organs continue to grow, while the number of donors does not. As a result, patients are arriving at transplant in worse and worse shape, which compromises success of the transplant.

And we may have already optimized the benefits of immunosuppression. The major focus now is on learning to manipulate the immune system, particularly with dendritic cells and bone marrow, so that a potential recipient can be made immunologically tolerant of a prospective transplant. That would relieve the need for chemical immunosuppressants and eliminate the dangerous side effects such drugs induce—and, importantly, greatly improve the quality of life for transplant patients.

As an aside, we can point out that the challenge of histocompatibility is a major driving force in the new field of **therapeutic cloning**. The idea behind therapeutic cloning is to use a patient's own stem cells to rebuild a defective organ, rather than replacing it with a transplant.

At present one of the most promising ways of doing this would use **embryonic stem cells**, which we know have the potential to develop into any cell type in the body. Unfortunately, it is not possible to isolate embryonic stem (ES) cells from an adult; they are only found in embryos. Usual sources for ES cells are the embryos generated during in vitro fertilization but not used for reimplantation into the mother. But that means that the ES cells would come from essentially a random donor, with whatever HLA alleles the parents of that embryo had, and they would be rejected unless the immune system is strongly suppressed.

A way around this is something called **nuclear transplantation**, which requires first of all a woman willing to donate a human egg. The nucleus is removed from her egg, and a nucleus taken from a cell of the prospective recipient is inserted in its place. The egg is allowed to divide six or seven times, giving a cell mass of a hundred or so cells, from which ES cells can be isolated and grown in the lab. These cells could then be inserted into the recipient at the appropriate tissue site and used for regenerative purposes. Since these ES cells express only recipient HLA antigens, there is no possibility of rejection.

HOW ARE TRANSPLANTS ACTUALLY REJECTED?

The Discovery of Killer Lymphocytes

It was during the course of laboratory investigations into the immunology of organ transplantation that lymphocytes that could kill other cells were first discovered. A young physician-scientist from Belgium, Dr. André Govaerts, had come to the United States to study what was known about the new and still struggling field of organ transplantation. As part of his medical and research activities, he was looking at the rejection of kidneys transplanted between genetically nonidentical dogs. By the time he was doing his experiments, it had generally become accepted that transplant rejection was caused by lymphocytes, rather than antibodies, although how cells did this was completely unknown.

At any rate, knowing that lymphocytes were somehow the cause of rejection, he decided to see what he could observe in the microscope. He transplanted a kidney from dog B into dog A, keeping dog B alive and healthy. After A had rejected B's kidney, he snipped out some of B's connective tissue cells and grew them in a Petri dish in an incubator at body temperature. The cells attached to the surface of the plate, spreading out and starting to grow in what cell biologists refer to as a **monolayer**—a single layer of cells whose edges all touch one or more adjacent cells across the dish.

After the monolayer was nicely formed and the cells seemed happy, he collected lymphocytes from dog A's lymphatic circulation—a large lymph collecting vessel in the dog's chest. He washed these, seeded them onto the monolayer of dog B's cells, and put the Petri dish back in the incubator.

Two days later, he was able to observe small holes in the mono-layer, always with one or more of A's white cells in the middle. A few days later, the monolayer was completely destroyed. But when A's white cells were placed on a monolayer of cells from dog C (genetically unrelated to B), nothing happened. The destruction of monolayer cells was antigen specific. He repeated this experiment several times, and always observed the same thing.

In 1960 he published a seminal paper describing what he had seen. But as a measure of how strong traditional thinking is, Govaerts did not title his paper something like "Killer Lympho-cytes—A New Mechanism of Immunological Defense." Rather, still unable to shake 70 years of immunological tradition, his paper bore the title "Cellular Antibodies in Kidney Homotransplantation."

It apparently seemed impossible to imagine that antibodies were not involved. He and others thought antibodies must be somehow sticking to the surface of the killer cells, possibly binding comple-ment and punching a hole in the graft cell membrane. But it was quickly shown that complement played no role in graft rejection.

So a new field—**cell-mediated cytotoxicity**—was born from this paper. The ability to monitor both recipient immune cells and donor graft cells as the destruction process took place outside the body (**in vitro**) made it possible to probe this process in great de-tail. Immunologists would spend the next 20 years trying to chase down how these "killer lymphocytes" did their work. As we saw earlier, this ultimately led to the discovery of at least two systems used by CD8 cells (perforin and Fas) for inflicting a "lethal hit" on recognized target cells.

The Role of "Passenger Cells" in Rejection

While some immunologists were trying to uncover the killing mechanism, others tried to understand how recipient T cells be-came activated against a transplant in the first place. One puzzle that presented itself early on was that although all of the cells com-ing in with a transplanted organ or tissue prominently display

MHC antigens, T cells will not respond at all to these antigens, with one exception. They will respond, and quite vigorously, to white blood cells from a foreign donor.

The conclusion from this was that when a transplant comes into the body, it is not the kidney or the heart or the lung itself that triggers a rejection response, but rather what are called **passenger leukocytes**. (*Leukocytes* is another name for white blood cells.) Some of these will be found in the blood vessels that come in with any transplant, but many are also scattered throughout the tissue of the graft itself.

As with so many other immune phenomena in which T cells play a key role, the trail led eventually to dendritic cells. Dendritic cells are absolutely key to initiating the T-cell response to a foreign transplant, and we are now quite sure that the critical passenger leukocyte for triggering rejection is the dendritic cell, which provides not only class I and class II MHC signals to virgin CD8 cells and CD4 helper cells, respectively, but also numerous cytokines and other signals required for T-cell activation.

Dendritic cells are found in virtually every tissue in the body, including lymphoid tissue, and they circulate in blood and particularly in lymph. If dendritic cells are destroyed in grafts prior to transplantation, rejection can be prevented. Unfortunately, it is not practical to do this on the scale needed for most organ transplants.

So how do host T cells encounter graft dendritic cells when a transplant first comes into the body? In our bodies, our own dendritic cells traffic back and forth between tissue sites and lymph nodes, usually traveling in the lymph fluids. This appears to be the case when an organ is transplanted into another person. Not knowing they are in a foreign body, the passenger dendritic cells migrate out of the transplant, slip into the lymph fluid, and drift away to the nearest lymph node. Do they have a surprise waiting for *them*! It's like a wasp that accidentally strays into a bees' nest.

Once inside the node, they encounter host CD4 and CD8 T cells that immediately go into a state of high activation, recognizing the dendritic cell as foreign and destroying it. The T cells then head out into the lymph and blood to look for the source of the "intruders." When they enter the blood vessels of the transplanted organ, which were connected to the host circulation by the transplant surgeon, they find themselves surrounded by wall-to-wall intruders!

The first foreign MHC products they see are on the walls of the blood vessels themselves. The T cells begin destroying the blood vessels and then enter the surrounding tissues. Because they were previously activated by donor dendritic cells, they can now attack and destroy nonleukocyte cells of the donor organ as well. But in fact, it is quite likely that the fatal blow was already delivered when the T cells attacked the transplant vasculature; once the blood supply to an organ is disrupted, it will quickly die from a lack of food and oxygen. The armies of macrophages that follow the T cells into the graft make short work of the debris left over from the attack.

BUT IS IT KILLER CELLS, OR IS IT INFLAMMATION?

Ever since André Govaerts discovered killer lymphocytes, it has been assumed that cell-mediated cytotoxicity by CD8 T cells could explain graft rejection completely—and indeed it *could*, at least theoretically. If the individual cells of a graft, or at least individual cells of the graft vasculature, are attacked and killed by CD8 cells, then clearly the graft will die.

But for many years, some researchers kept reminding their colleagues that graft rejection reactions in the body (**in vivo**) are accompanied by an intense inflammatory reaction, starting within just a day or two of the transplant. We have seen the collateral damage that can be done to otherwise healthy tissues during inflammation. Might that be sufficient to explain graft rejection? Surely it must at least be a factor?

In fact, for many years after killer cells were described, medical textbooks still listed graft rejection as simply another type of hypersensitivity reaction, specifically, delayed-type hypersensitivity (DTH), the basis of the tuberculin reaction and poison ivy reactions. It was viewed as largely an inflammatory process, driven in part by antigen-specific CD4 and CD8 cells that recognized the graft MHC as foreign and released inflammatory cytokines. The killing reaction might get a line or two at best in the overall description, but the major mechanism at play, the textbooks implied, was inflammation. The resulting damage was assumed to be from innate immune mechanisms rather than the T cells themselves.

Of course, this mightily annoyed the segment of the immunology research community that had devoted entire professional lives to the study of CD8 T-cell–mediated killing. Surely cell-mediated cytotoxicity must be more than just a footnote? So they devised experiments to determine the respective roles of direct graft killing by CD8 cells versus inflammation.

One of the earliest of these experiments was carried out in Sweden in 1972. Sarcoma cells from two inbred mouse strains, A and B, were mixed together in equal parts and implanted under the skin of an A mouse. The rationale for the experiment was this: the A mouse will recognize the B sarcoma cells as foreign and destroy them—not because they are tumor cells, but because they are a foreign transplant. But the A mouse will not mount an immune response to the A sarcoma cells because they are self.

Now, if the mechanism of killing is inflammation, the effects of which (toxic cytokines, generalized phagocytosis) are nonspecific, then both tumors should be swept away by the inflammatory response, because the A and B cells are packed cheek by jowl together at the same subdermal site. But if the mechanism of rejection is antigen-specific killing by CD8 cells, then only the B cells should be killed.

The result was very clear. A tumor did grow out of the site, but all of its cells were of A type. Only the B tumor cells had been killed—all of them. Had even one been left behind, it would have grown out and contributed to a mixed A/B tumor. This was a very strong argument against a generalized, nonspecific mechanism of killing. This experiment was repeated many times by others, using rather sophisticated variations, but always with the same result. Graft rejection was exquisitely specific; only those cells directly recognized by CD8 cells were killed. Cells immediately adjacent to them were not. This was seen as incompatible with a DTH-instigated inflammatory reaction.

Well, you would think those favoring direct CD8 cell killing as the mechanism of graft rejection would have been content to leave it at that. But no, they had to take it one step further. In the 1990s, it became possible to create something called a **gene knockout mouse**. This is done using embryonic stem cells. By appropriate manipulations, it is possible to derive an entire mouse from a single embryonic stem cell. And while the stem cells are growing in the lab, it is possible to alter them genetically. It is possible, for example, to remove a particular gene and see what happens. The resultant mouse will lack this gene in every cell of its body. This is a great technique for figuring out exactly what a particular gene does.

When knockout mice were created that lacked the **perforin** gene, and hence had no perforin in their CD8 cells, the CD8 cells lost their ability to kill graft cells in vitro. The researchers then looked at the ability of the perforinless mice to reject skin allografts. They rejected the grafts as rapidly as mice that still had perforin. The possibility that the second killing mechanism, Fas, could be responsible for rejection was quickly eliminated. Suddenly you could hear a giant "huh?" echoing from coast to coast, followed by much weeping and gnashing of teeth.

Transplant immunologists are still trying to unravel what this means. Does it mean that the direct, CD8-mediated killing that is so spectacularly evident against graft cells in vitro plays no role in graft rejection in vivo? That's what the data say; the data say that in the absence of these two killing mechanisms, graft rejection will still occur.

Obviously there are other mechanisms involved that also cause graft rejection. We saw in the last chapter that interferon-γ released by CD8 cells can compromise blood supply in tumors, leading to their failure to grow. Most transplants are brought in with their own blood vessels intact, so it is unclear if interference with blood supply could be a mechanism in graft rejection. But since we don't know the answer at present, everything is on the table. The situation in viral infections, which is presumably what CD8 killing evolved to deal with, is more clear. There, if perforin is absent, mice infected with many viruses cannot overcome their infections. And as we would expect, it is just those viruses that live inside cells most of their lives that cannot be eliminated in the absence of CD8 cell killing.

Transplant surgeons (as opposed to transplant immunologists, the researchers who have to figure out what is going on) don't really care which it is—inflammation or direct CD8 killing. All they care to know is that whatever the mechanism, T cells cause it, so they just focus on interfering with T-cell activation or T-cell function as a means of preventing rejection. And that seems to work pretty well.

But it is a bit ironic that the system in which CD8 killer cells were first discovered—rejection of organ transplants—now seems to be the system where the direct killing function of these cells may be the least important.

14

First Defense

THE IMMUNE SYSTEM AND BIOTERRORISM

Bioterrorism is the use of biological organisms or their derivatives to sow terror in a civilian population. Bioterrorism is an offshoot of biological warfare, and like most progeny it differs from its parent. The main difference is that biological warfare is a highly organized aggressive activity carried out by one state against another, usually through a military arm, with the sole aim of killing or disabling people. Bioterrorism, while using many of the same agents and tactics as biological warfare, is a more ad hoc activity carried out by individuals or political groups against other political groups or states, with a mixture of objectives.

Biological warfare itself has a long if occasionally crude history, including dipping arrowheads and spear points into rotting cadavers or feces, or lobbing entire diseased corpses over town or castle walls. The perpetrators obviously had little understanding of what they were doing, so it may be less than accurate to call this biological warfare.

But once the basis for infectious diseases was uncovered in the second half of the nineteenth century, it didn't take long before biological warfare became a highly precise science. By World War I, and on through World War II, virtually every major world power had established scientific research units dedicated to the subject. In the United States the War Department (precursor of today's Department of Defense) established a special biological warfare facility at Fort Detrick, Maryland. Anthrax and plague were among the microbes of choice in the programs of most countries.

However, with the exception of Japan during its occupation of China and Manchuria, there was no extensive use of biological agents against either military or civilian targets during World War II. President Nixon ended the active development of biological warfare agents in the United States in 1969. And finally, in 1972, over 100 countries signed a Biological Weapons Convention that outlawed biological weapons and mandated destruction of existing weapon stockpiles.

Bioterrorism has a more limited history. The first documented instance of bioterrorism in the United States was carried out by an Oregon cult (the Rajneeshees) in 1984, in an attempt to manipulate a local election. Over 700 people were made ill with *Salmonella* bacteria, though none died. In the early 1990s, the Japanese Aum Shinrikyo cult released anthrax spores in several Japanese urban settings, with fortunately few casualties. Shortly after the September 11 attacks in the United States, unknown individuals used the Postal Service to disseminate anthrax spores in letters.

The FBI had uncovered and foiled two additional bioterrorist plots before then. In 1992, an antigovernment group in Minnesota—the "Patriot's Council"—planned to use a toxic extract of castor beans called **ricin** to kill local and federal law enforcement personnel. In 1995, a member of a white supremacy group was arrested and sentenced to 18 months' probation for stockpiling bacteria that cause the plague. He could produce no legitimate reason for possessing such quantities of a deadly pathogen. So bioterrorism has already arrived on American shores, and the enemy, so far, is us.

In the United States, the Centers for Disease Control (CDC), in Atlanta, Georgia, is the primary federal agency responsible for coordinating all scientific, medical, and public health aspects of the federal response to potential and actual bioterrorism. In 1999, even before the September 11 attacks on the World Trade Center and the almost immediately subsequent anthrax scares, the CDC commissioned a detailed study of agents that could be used in bioterrorist attacks. Those that proved to be of the greatest concern, based on factors such as lethality, ease of dissemination, and ability to induce panic and social disruption, were given a "Category A" designation (Table 14.1).

Federal, state, and local governments have a wide range of programs poised to be activated at the first hint of a bioterrorism attack, including rapid identification of the biological agents involved, tracking and containing their spread, and identifying and treating affected individuals. That is all well and good, and would doubtless greatly reduce the potential damage from a bioterrorism attack.

But in the early stages of any such attack, your primary—your only—defense will be your own immune system, honed as we have seen over millions of years of evolutionary selection to respond rapidly and effectively to invasion of your body by potential microbial predators and their toxins. The microbes around which a bioterrorism attack could be mounted will likely be selected in part on the basis of a known poor immune response by humans to the agent involved as well as maximum debilitation—or panic—caused by the attack.

In the sections that follow we take a look at the microbes and toxins currently deemed by the CDC as most likely to be used as bioterrorism agents.

ANTHRAX

Anthrax is a disease caused by the bacterium *Bacillus anthracis*. It affects animals, mostly grazing herbivores such as sheep, goats, and cows. Humans are vulnerable to anthrax infection, but we have learned over the centuries how to avoid it, and veterinarians are skilled at keeping domestic livestock free of the disease. Fewer than 250 cases of naturally acquired anthrax in humans have been reported in the past 50 years in the United States. In less developed parts of the world, annual new cases of anthrax are considerably more.

Anthrax is arguably the most serious threat on the CDC's list of Category A bioterrorism agents. It is deadly: the mortality rate for

TABLE 14.1

Biological Agents Classified by the Centers for Disease Control as Category A Potential Bioterrorism Agents

Disease	Agent	Lethality	Treatment	Vaccines
Anthrax	*Bacillus anthracis*	High	Antibiotics	Yes
Smallpox	*Variola major*	Moderate-high	None	Yes
Plague	*Yersinia Pestis*	High	Antibiotics, antiserum	(Yes[b])
Tularemia	*Francisella tularensis*	Low-moderate	Antibiotics	(Yes)
Botulism	*Clostridium botulinum*	High	Antiserum	(Yes)
Viral hemorrhagic fevers	Ebola, Marburg[a] viruses	High	None	(Yes)

Among the criteria for classification as a Category A agent: high level of virulence or toxicity in humans, feasibility of large-scale production and dissemination as an aerosol, readily spread from person to person, lack of effective treatment and of public health preparedness, and potential for public panic and social and/or economic disruption. Category B agents include (but are not limited to) ricin toxin, Staphylococcus enterotoxin B, and encephalomyelitis virus. Category C agents include such things as hantavirus and multidrug-resistant tuberculosis.

[a] Other hemorrhagic fever viruses include Lassa fever virus, various arenaviruses, Rift Valley fever virus, yellow fever virus, Onsk hemorrhagic virus, and Kayasanur virus.

[b] This vaccine is not effective against pneumonic plague, however. Vaccines for this form of the plague are nearing readiness for clinical trials.

untreated anthrax can range from 20% to 100%, depending on the form of infection (see below). The World Health Organization (WHO) has estimated that 50 kg (about 110 pounds) of anthrax spores released in the air over a population of 5 million could result in serious disease in 250,000 people and lead to as many as 100,000 deaths, putting such an incident on par with a nuclear bomb attack! There is already substantial public awareness of just how deadly anthrax is, and news of an anthrax attack in a dense urban area would doubtless create major panic and civil disruption, a major aim of any terrorist attack.

Part of what makes anthrax so deadly is that *B. anthracis* forms spores. Most bacteria, when they run out of food, simply starve to death. A few bacteria, however, are able to enter a state of suspended animation—to convert to **bacterial spores**. Spores do not carry out any metabolism, do not need water, and are extremely resistant to heat and many toxic chemicals. Properly prepared, they are hard, dry particles easily carried on wind, which makes them perfect for use as bioterrorism agents. When they land on a surface possessing moisture and nutrients—human skin or lungs, for example—they rapidly revert from spores to normal bacterial cells in a process called **germination**. Spores can survive in their dehydrated state for several decades.

Anthrax spores can be inhaled or can settle on exposed areas of skin. Both would likely occur in most exposed individuals, and both pathways of entry can result in disease. In **inhalational anthrax**, many of the spores are engulfed by lung macrophages. Initial symptoms are fever, achiness, and often a sore throat. Many of the ingested spores are able to germinate inside the macrophages, eventually destroying them and escaping into surrounding tissues. Other spores will settle directly on soft, wet lung tissue, germinate, and begin to divide. In either case, actively dividing bacteria quickly migrate through lymph and blood to other parts of the body. It doesn't take long before rabidly dividing, healthy bacteria have spread everywhere. Mortality in untreated inhalational anthrax can approach 100%.

Spores settling on the skin (**cutaneous anthrax**) can enter the body through cuts or abrasions. Once inside, they follow a similar path. Some germinate locally and cause redness and itching that can develop into local skin ulcers (Figure 14.1); many will spread to other parts of the body, germinating as they go. However, death from this form of anthrax, untreated, rarely exceeds 25%.

Anthrax infections can be treated with the antibiotic Cipro. Effectiveness depends on the form of infection (it is most effective against cutaneous anthrax) and how long the infection has been in progress. There are no known cases of transmission of anthrax from one human being to another, an important factor in the management of an anthrax attack.

B. anthracis produces two deadly toxins that are responsible for the illness and death accompanying anthrax infections, regardless of the mode of entry. **Edema toxin** causes water to escape from host cells in the vicinity of anthrax bacteria, causing massive swelling, which interferes with normal tissue functions. **Lethal toxin** cripples the innate, and thus the adaptive as well, immune responses, al-

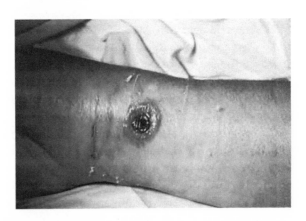

FIGURE 14.1
Skin ulcer resulting from a cutaneous anthrax infection. (*Courtesy NIAID Biodefense Image Library.*)

lowing unimpeded replication of anthrax bacteria in the body. The resulting inflammation and accumulation of fluids and bacterial byproducts lead to rapid deterioration of host metabolic functions, profound shock, and death. After consuming whatever is left of the host and running out of food, the bacteria generate more spores and drift away on the wind looking for a new hotel. Cremation is recommended for any remains.

There have been few studies of the immune response to anthrax in humans, because natural infections are now so rare. Most of what we know about the immune response comes from studying infections in animals and human responses to anthrax vaccines. The most important response is the production of antibodies against the two anthrax toxins. These antibodies block the ability of the toxins to bind to cells and also "tag" both spores and bacteria for removal by macrophages. CD4 T cells necessary to help B cells make these antibodies are important, but since *B. anthracis* does not live inside cells (aside from their brief transit through macrophages), CD8 T cells probably play little role in immune defense.

Unfortunately, useful levels of antibodies rarely develop in natural infections, because the toxins so quickly knock out the key cells involved in starting an immune response. The loss of dendritic cells in particular, so crucial in triggering inflammation and activating T-helper cells, is perhaps the most serious damage caused by anthrax toxins.

Several vaccines against anthrax toxins work reasonably well to induce antitoxin antibodies. Unfortunately, all of the current vaccines require several injections over at least several weeks, and so would be of little use for the early victims of an anthrax attack. They also have uncomfortable side effects for some individuals.

However, people in the immediate vicinity of an attack would likely be given these vaccines anyway, since anthrax spores will be everywhere and can persist for decades. Researchers are working on improved vaccines, some based on the DNA technology discussed in chapter 7, that could generate protection much more quickly and with fewer side effects. Such a vaccine could be of more use during an actual attack, but for economic and practical reasons would probably not be used to immunize the general population in the absence of an attack.

The CDC has recommended that we might revert to one of the oldest forms of immunization—passive immunization (chapter 2)—for anthrax. This involves the injection into one individual of antibodies made in another individual. Five of the 11 victims of the postal anthrax attack in 2001 died despite intense antibiotic treatment. The U.S. Department of Heath and Human Services (HHS) is currently contracting with several private companies for production of such antibodies. Passively transferred antibodies might work even more rapidly and effectively than antibiotics. The antibodies would be directed at the anthrax toxins and block them from attacking host cells. But passive immunization is intended only for the immediate treatment of victims, not for preventive immunizations of entire populations.

SMALLPOX

Today it is hard to imagine that smallpox was once one of the deadliest diseases on this planet, probably exceeding even the plague in the cumulative number of people killed throughout history. When contracted through the lungs by breathing in air into which an infected person had sneezed or coughed, it routinely killed 20% to 30% of unvaccinated individuals, well into the twentieth century, and left the rest badly disfigured for life. Smallpox can also be spread by person-to-person physical contact, although the resultant disease is usually less fatal.

Smallpox is caused by an orthopoxvirus called *Variola major*. Almost uniquely among human pathogens, *V. major* has no known animal or insect **reservoir** (a host in which it can reproduce without causing disease, or at least death). In its present form, it appears to be entirely dependent on human beings for its propagation and survival.

Smallpox is the first disease-causing microbe to be purged from the human species, by a worldwide immunization campaign launched by the WHO in 1967. By 1972, routine vaccinations were discontinued in the United States because of a small risk of active disease from the vaccine itself. The fact that *V. major* could not retreat into an animal reservoir during this campaign was a major factor in its eradication. Today *V. major* officially exists only as frozen stockpiles at the CDC in Atlanta, and in a former biological warfare research center near Novosibirsk, in Russia.

Smallpox is on the CDC Category A list because of its high mortality rate and because, as a viral disease, it is essentially untreatable. It spreads very efficiently as an aerosol, and the virus is relatively stable. Also, like anthrax, there is enough residual public awareness of the deadliness of smallpox that news of its spread in a terrorist attack would likely generate considerable panic and social disruption. Since for the past 30 years almost no one in this country has been vaccinated against smallpox, the U.S. population is highly vulnerable to this disease.

Smallpox was used as a weapon by the British in the French and Indian Wars (1754–1767). Blankets that had been used to wrap infected British soldiers were distributed to Indian tribes cooperating with the French. Although this means of spreading smallpox is less deadly for the initial victims, they in turn spread it as an aerosol through coughing and sneezing. The overall fatality rate among the Indians was well over 50%.

In the course of a *V. major* infection, viruses settle into airway tissues and are swept along into regional lymph nodes where they provoke an immediate response by the innate immune defense system. This results in some combination of mild fever, chills, and achiness. When the virus reaches the skin (from the inside out, as it were), a rash appears, followed by the formation of multiple, closely packed blisters on all parts of the body, but particularly the face and neck. These blisters also form in the mouth and throat, where they break easily, dumping their viral load into the saliva. This aids in the further spread of the virus into the general population through coughing and sneezing. The cause of death, in those cases that are fatal, is unclear but may be due to the enormous buildup of antigen–antibody complexes, which trigger rampant inflammation and tissue damage in kidneys and lungs.

As we saw in chapter 5, T cells are a major part of the immune defense against many viruses. Because smallpox was largely eradicated in humans by the early 1970s, when T cells were just beginning to be studied, we know almost nothing about T-cell immunity against smallpox. It would seem likely that CD8 killer T cells play a role in controlling human smallpox infections. But techniques for studying CD8 T cells in humans were not worked out until the mid-1970s. Moreover, our views of innate immune mechanisms, and their interaction with the adaptive immune system in the activation of T cells, have changed radically in the last 10 years and have never been examined in smallpox infections. Natural killer (NK) cells were not even discovered until 1975. The lack of an animal model for studying smallpox has long been a major drawback. Recently, however, *V. major* has been used to produce infections in macaque monkeys, and information about how this virus interacts with their immune systems may be useful in designing smallpox vaccines and antiviral drugs for smallpox infections in humans.

The few insights we do have into the possible course of the human cellular immune responses to *V. major* come from studies in the late 1970s on volunteers receiving smallpox vaccinations. Immunizations for smallpox over the years have never been carried out with *V. major*—it is too deadly—but rather with a closely related orthopoxvirus called **vaccinia**. The exact origins of this virus are unclear (chapter 7). Vaccinia is injected in a fully viable form. It induces a mild local reaction at the site of injection that usually resolves in 7 to 10 days. Protection from subsequent infection by *V. major* after vaccinia immunization is excellent, but about 1.6 cases per 1 million immunizations progressed from mild reaction to more serious disease, and occasional deaths, which is why vaccination for smallpox was abandoned in 1972.

Nevertheless, a number of healthy individuals agreed to serve as volunteers in a study published in the late 1970s. In this study, both mice and humans produced killer cells against vaccinia-infected target cells. The killer cells from mice were clearly CD8 killer cells, but the killers produced by the human volunteers were not. The best guess at the time was that NK cells, and possibly neutrophils, were involved in killing vaccinia-infected cells in humans. Later studies suggested that CD8 killers are in fact produced at low levels in response to vaccinia. At any rate, we cannot be sure that the immune response in humans to infection by *V. major*, the actual pathogen in smallpox, would be exactly the same as that produced following immunization by vaccinia. For designing future vaccines, we would really like to know this. But it seems unlikely we ever will.

We do know that the amount of virus-neutralizing antibody produced in response to immunization with vaccinia directly correlates with subsequent protection to *V. major*. During a natural infection with *V. major*, antibody production peaks after about three weeks and remains high for several years. Most studies also suggest a role for vaccinia-induced antibody in tagging *V. major* for phagocytosis and destruction by macrophages and neutrophils. The Department of Health and Human Services is currently funding development of so-called third-generation vaccinia-based vaccines, specifically in response to concerns about bioterrorist attacks using smallpox.

PLAGUE

References to the plague in human history date back to at least 500 B.C., although we don't really know if that plague was the same as the three pandemics that swept Europe and Asia in the Middle Ages, which is what we recognize as plague today. There were several major and many minor pandemics in Europe in the fourteenth through eighteenth centuries. They were deadly. Although we have no precise figures, as many as 200 million people are estimated to have died.

The non–spore-forming bacterium *Yersinia pestis* has been associated with the European outbreaks, and *Y. pestis* DNA has actually been extracted from dental remains found in graves of persons dying from the plague. We still see occasional incidents of *Y. pestis* plague, with several thousand new cases arising annually throughout the world. There have been about 400 cases in the United States since 1950, mostly in the southwest.

Y. pestis can cause several types of disease in humans, depending on how the infection is acquired. **Bubonic plague** results when a human is bitten by an insect, usually a flea, carrying *Y. pestis*, which it acquired from previously biting an infected animal. In urban areas, the most common animal carriers are rats and squirrels and the occasional house cat. *Y. Pestis* can also jump from animal fleas into fleas more at home in humans, which greatly aids human-to-human spread of the disease. In the middle ages, most people had fleas in abundance. Prior to the introduction of antibiotics in the 1940s and 1950s, fatality rates of 50% or more for bubonic plague were not uncommon.

After transmission through a bite, *Yersinia* bacteria begin to replicate and are swept along to nearby lymph nodes in lymph fluid. At various stages in this journey they may be engulfed by macrophages, but are relatively resistant to being digested by them. A few days later the typical symptoms of a microbial infection set in: fever, chills, and general achiness, byproducts of activation of the innate immune system.

As the bacteria continue to replicate inside the lymph nodes, the nodes become greatly enlarged (**"buboes"**) and very tender. They can grow as much as four inches across. If the bite occurs in the lower part of the body, lymph nodes in the groin are preferentially affected. Bites in the upper body regions tend to deliver bacteria to lymph nodes in the armpit and neck. Untreated mortality rates are around 50% of those infected.

If the bacteria spill out of the lymph nodes and enter the general blood circulation, numerous other tissue compartments become involved and the infection is even more lethal ("septicemic plague"). Blood vessels are destroyed, resulting in gangrene in the extremities. This is probably the origin of the term "Black Death" for plague. Prolonged infection can also trigger shock, a common cause of plague death. If the bacteria invade the lungs (secondary pneumonic plague), the infection is almost always fatal and the bacteria spread more readily from person to person through sneezing and coughing.

Primary pneumonic plague, the second major form of plague, is particularly deadly. It occurs when *Y. pestis* is taken in directly through the respiratory system as opposed to an insect bite. Untreated mortality rates approach 100%. Symptoms set in within a day or two after inhalation of infectious *Y. pestis* and are initially indistinguishable from other forms of aggressive pneumonia. Aerosolized *Y. pestis* and the pneumonic plague that results would likely be the choice of terrorists. Bubonic plague carried by fleas is difficult to spread over a large area and does not pass easily from person to person. Experience in diagnosing and treating pneumonic plague is very limited in the United States. Moreover, many currently used antibiotics have never really been tested against *Y. pestis* in humans.

The only modern-day use of plague for biological warfare was by the Japanese during occupation of China in World War II, when they released plague-infected fleas onto civilian populations. Both the United States and the Soviet Union pursued development of aerosolized *Y. pestis*, but these appear never to have been used. Aerosols of course would induce pneumonic plague and could be unbelievably deadly. The WHO estimates that 50 kg of aerosolized *Y. pestis* spread over an urban population of 5 million would infect at least 150,000 people, causing at least 36,000 deaths.

There have been no documented attempts to use *Y. pestis* as a bioterrorism agent. However, in 1995, a microbiologist was arrested for fraudulently obtaining large amounts of plague bacteria, with no obvious legitimate scientific purpose. And in 2004, a respected physician-scientist at Texas Tech University was sentenced to two years in prison for grossly mishandling and illegally shipping to Tanzania vials containing infectious *Y. pestis*—on a commercial airliner, no less! No connection with bioterrorism was alleged or proved.

We know few details of the human immune response to *Y. pestis*, especially primary pneumonic plague infections, because of the scarcity of cases. Most of our recent insights into human immune responses have never been examined in plague patients. The standard vaccine for many years, based on a whole-cell, killed form of *Y. pestis*, is no longer used. A second vaccine based on a live but attenuated form of the bacterium induces good protection against bubonic plague but unfortunately little or no protection against pneumonic plague. Producing an effective vaccine for pneumonic plague is an active area of research, driven largely by concern about use of plague for terrorism.

From animal studies we know that in bubonic plague (and we assume in pneumonic plague as well), plague bacteria quickly make contact with host macrophages, dendritic cells, and neutrophils. But immediately upon contact, the bacteria inject substances into the cells that greatly reduce their phagocytic function. Even for those bacilli that are successfully engulfed, additional substances released inside the cell inhibit degradation of the bacilli and allow them to replicate. Another chemical released while the bacilli are still outside the cell kills off nearby NK cells, a vital component of the early response to viruses. So right off the bat, crucial elements of the innate immune system that otherwise would ordinarily slow the infection, and help kick off an adaptive response, are seriously disabled.

The B-cell response to *Y. pestis* requires CD4 T-cell help, and CD4 T cells also produce numerous cytokines that help fight the infection. Little is known about the role of CD8 cells in the response. Many *Y. pestis* bacteria remain extracellular during an infection, but many also manage to replicate within macrophages and perhaps dendritic cells. Whether these intracellular bacteria elicit a protective CD8 response is not known. Many people die without ever mounting an effective antibody or T-cell response

As noted above, currently available vaccines induce a good antibody response that can control bubonic plague but offer little protection against pneumonic plague, which we expect would be the form we would encounter in a terrorist attack. Moreover, these vaccines require multiple injections to be effective, do not give long-lasting immunity without booster shots, and can have numerous undesirable side effects. There is currently an intensive laboratory campaign to develop vaccines effective against pneumonic plague. Hopefully one can be found that acts quickly enough to be of some use during an attack while having acceptable side effects. Several are based on recombinant DNA. Tests on animals look promising, and one of these has recently received Food and Drug Administration (FDA) fast-track approval for human clinical trials.

BOTULISM

Botulism is caused by a protein toxin released by several strains of bacteria of the genus *Clostridium*, including the eponymous *Clostridium botulinum*. This toxin is the most lethal biological poison known—100,000 times more poisonous than sarin gas. One gram (about 1/28 of an ounce) in aerosol form could theoretically kill 1 million people.

Unlike other A-list bacterial agents, such as anthrax, plague, and tularemia, where the agent is the bacterium itself, in the case of *C. botulinum*, we cut right to the chase—the isolated, purified toxin produced by the bacterium, which is entirely responsible for the disease caused by this bacterium, is the agent. Like *B. anthracis, C. botulinum* is a spore-forming bacterium, and could readily be aerosolized as such. But since the object of delivering the bacterium is to deliver its toxin, it is more efficient to simply aerosolize the toxin itself, which is relatively easy to collect and concentrate.

The toxin produced by *C. botulinum* and related strains is a **neurotoxin** that prevents the brain from telling muscles to contract. Individuals poisoned by botulinum toxin through food experience extreme muscular weakness and have difficulty seeing, speaking, and swallowing. Their brain and associated mental functions are not impaired, but their muscles just cannot do work, including the muscles that control breathing. Death comes mostly from respiratory failure.

Intact *C. botulinum* bacteria were fed by the Japanese to prisoners during their occupation of Manchuria in World War II. The results, as far as they are known, were uniformly lethal. The Aum Shinrikyo cult in Japan attempted to carry out attacks in Tokyo in the 1990s using botulinum toxin, but for technical reasons were unsuccessful. Research on botulinum toxin as a possible biological weapon was carried out by the United States and other countries over the years but never used. United Nations inspectors determined during the 1990s that Iraq had prepared about 5,000 gallons of concentrated toxin, some of which was found by inspectors to have been loaded onto missiles, ready for use. How effective these would have been as bioweapons is, however, unknown.

Natural infection by *C. botulinum* can occur by eating contaminated food, usually vegetables, though the name botulinum in fact derives from the Latin for sausage (*botulus*), a common food contaminated by *C. botulinum* in former times. The poisonous effect of food contaminated with this bacterium is due entirely to the toxin, which it readily secretes into its surroundings. *C. botulinum* can also enter through wounds, for example, in the foot, when walking in soil harboring this bacterium. It can also infect needle puncture wounds and can be a problem among intravenous drug users. The toxin itself will not penetrate unbroken skin.

Aerosolized toxin is rare in nature, but has been prepared in numerous laboratories. Whatever the mode of entry into the body, the toxin quickly gains access to the blood and lymph, from where it reaches the points at which nerves make contact with muscles **(neuromuscular junctions)**. With the purified toxin there is no response by the innate immune system and none of the signs associated with a microbial infection (fever, chills, achiness, etc.). Common early signs of poisoning, which begin 12 to 72 hours after ingestion (depending on dose), include difficulty in speaking and swallowing, dry mouth, blurred vision, and extreme muscular weakness. Treatment almost always requires extended stays in intensive care units, which in the event of a large-scale attack would rapidly overwhelm the capacities of even the best hospital systems.

A complicating factor with botulin toxin is that it has a number of useful medical applications when delivered in extremely low doses and in a highly targeted fashion. A number of disorders are characterized by excessive or involuntary muscle contractions, and botulin toxin (**"Botox"**) can be used to alleviate these spasms. The list of such disorders is quite long and includes many genuinely serious conditions. Thus, plans for general immunization of large populations could interfere with the use of Botox for therapeutic purposes. Botox has also been used in recent years for cosmetic purposes, although this is obviously not a major public health consideration.

The immune response to botulin toxin in humans is not well understood, again because of the scarcity of cases of natural infection available for study. Anyone exposed to botulin toxin must immediately undergo intensive treatment and is not an appropriate subject for being poked in his or her lymph nodes. So most of what we know about the immune response to the toxin comes from studying the response in animals to C. *botulinum* and its toxin, or the human response to **botulin toxoid**, a form of the toxin that has been crippled in its ability to cause disease without altering its ability to induce an immune response. Botulin toxoid indeed does induce a good immune response in humans, but to what extent this mirrors the response to native botulin toxin is not known.

Botulin toxin as a bioterrorism agent would involve purified protein with no contamination by viable microbes. So, as noted earlier, there would be no signs of an active microbial infection to signal its presence—no fever, no chills, no achiness. Moreover, the dose received by any individual would likely be extremely minute. Given its incredibly poisonous toxicity, such doses would be sufficient to cause severe disability or even death, but could be insufficient to trigger a meaningful immune response. About half of the botulism patients who have had their blood analyzed do show antitoxin antibodies, but the response is not particularly strong and the majority of these patients who survive do not go on to develop an immunological memory of the toxin. This latter is attributed to the low levels of botulin toxin in their system.

The only current botulin toxin vaccine for use in humans in the United States was first produced in 1970. It consists of a mixture of various forms of the toxin that have been chemically treated to produce toxoids. Reasonable levels of antibody capable of neutralizing botulin toxin have been induced in volunteers after two to three injections. It is intended only for use in vaccinating military personnel in the event of a threat of biological warfare and people working in laboratories where botulin toxin is studied. A number of DNA vaccines, in which genes for various portions of the toxin known to trigger antibody production are introduced into individual subjects (Chapter 7), have shown great promise in animal studies. Recently, one such vaccine received clearance from the FDA for fast-track human clinical trials.

At present the only therapy for botulism is injection of toxin antibodies (antitoxin) produced in horses. This is not without side effects, mostly due to the fact that the horse antibodies are recognized as foreign by the human immune system, so multiple administrations of the antitoxin would not be possible. Researchers have focused on the production of toxin antibodies that are less likely to trigger an immune response in humans, but human trials for FDA approval lie some years in the future.

TULAREMIA

Tularemia is caused by the bacterium *Francisella tularensis*, named for its discoverer, Edward Francis, and the place of its discovery, in 1911, in Tulare County, California. It had been associated with a variety of plague-like diseases in animals such as deer-fly fever, rabbit fever, and tick fever, among others, all of which are now grouped as various forms of tularemia. Humans can be infected by contact with *F. tularensis* in the wild, although such incidents are relatively rare, and humans do not readily transmit the resulting infection to others. Contact can result from handling of infected animals or through insect bites, but these infections are usually mild.

The most serious incidents of tularemia in humans come from inhalation of bacteria, often through handling of contaminated hay or other grains. "Inhalation tularemia" requires only a few bacteria, whereas infection through other routes usually requires exposure to millions of bacteria. If *F. tularensis* were to be used as a bioterrorism agent, it would almost certainly be in aerosol form. Because of the low incidence of inhalation tularemia in the United States, a large outbreak of this disease in a concentrated area would lead to an immediate suspicion of bioterrorism.

Tularemia was investigated by several countries between 1930 and 1970 as a potential biological warfare agent. The bacterium can be concentrated into a paste, which can be freeze-dried and then milled into a fine powder suitable for distribution through the air. A WHO study estimated that 50 kg of bacteria (about 110 pounds) in aerosolized form, spread over a population of 5 million people, would incapacitate about 250,000 people and cause nearly 20,000 deaths. The United States retained stocks of *F. tularensis* through the late 1960s, but these were destroyed in the general obliteration of such stockpiles in the early 1970s. Current military research with this microbe is restricted to defensive strategies.

Inclusion of *F. tularensis* by the CDC as a Category A bioterrorism agent is due largely to its effectiveness when spread in aerosolized form. The initial signs of inhalational tularemia infection are no different from the signs accompanying most microbial infections. Most of what we know about the immune response to tularemia has been gleaned from studies of this disease in rodents. The course of infection and the resulting immune response in rats and mice appear to mimic closely the events occurring in human infection. *F. tularensis*, like *Mycobacterium tuberculosum* (chapter 6), invades and takes up residence in macrophages but manages to escape digestion and multiplies rapidly within the macrophage itself.

In the case of inhalation tularemia, the primary target is macrophages resident in the lungs, but the bacteria make their way to regional lymph nodes as well. The lung tissues become generally inflamed and can develop various forms of pharyngitis, bronchitis, and other forms of lung infection. One form or another of pneumonia is the most common cause of death in fatal cases. Tularemia is one of those diseases that have a "low index of suspicion" among doctors and laboratory personnel, which could also be a factor in its selection by bioterrorists.

Both innate and adaptive responses are mobilized in response to *F. tularensis*, but development of a vigorous adaptive response is absolutely essential to clearing an infection. Production of cytokines like IFN-γ and TNF-α are critical during the early innate immune response and are probably provided by dendritic cells, macrophages, and perhaps NK cells. Interestingly, neutrophils are able to scavenge and kill *F. tularensis*; apparently the tricks used by this bug in escaping lysosomal destruction in macrophages don't work in neutrophils.

As would be expected for an intracellular parasite, B cells and antibody play little role in the ensuing adaptive response. Effective, long-term immunity is provided almost entirely by T cells, both through enhanced production of IFN-γ and probably through direct T-cell–mediated killing as well, although there is little direct evidence for the latter in the current scientific literature.

There is a vaccine for tularemia, based on a live but relatively harmless strain of *F. tularensis*. But this vaccine is only marginally effective against inhalation tularemia, and it takes at least a week or two to build up a good level of protection after vaccination. In a bioterrorism attack using an aerosolized form of the bacterium, it is unlikely that this vaccine would be useful in protecting exposed individuals after the attack. Development of a more active vaccine, perhaps based on DNA (Chapter 7), should be a goal of those concerned with homeland security.

HEMORRHAGIC FEVER VIRUSES

Hemorrhagic fever viruses (HFVs) are by far the most deadly of human pathogenic microbes. The CDC has designated nine HFVs as potential bioterrorism agents (Table 14.1). We will focus here on only two of these, the **Ebola** and **Marburg** viruses (Figure 14.2). Both of these viruses are fairly recent additions to the repertoire of human pathogens, and not that much is known about their interaction with their human host. There have only been a dozen or so outbreaks of these viruses since their discovery in 1967 (Marburg) and 1976 (Ebola).

We do not know what animals serve as a reservoir for these viruses—hosts that harbor them without contracting serious disease. Several nonhuman primates, such as rhesus monkeys and macaques, are fully susceptible to the ravages of Marburg and Ebola infection and would be unlikely reservoirs. Cases of human infection tend to occur in clusters, the origins of which are not always clear; in several instances human infection seems likely to have originated from contact with infected monkeys. Once one

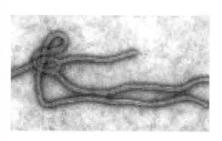 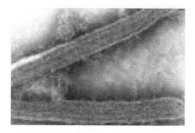

FIGURE 14.2
Ebola (left) and Marburg (right) virions. (*Courtesy NIAID Biodefense Image Library.*)

person has been infected, however, transmission to others occurs through contact with fluids or tissues from previously infected individuals.

Ebola and Marburg viruses would certainly fulfill the CDC requirement that an agent have the potential to cause "public panic and social disruption." Through books and films in the past two decades, plus regular media coverage of outbreaks, these viruses may have, along with anthrax, the highest public profile of the Category A agents. Because of the high mortality rate, the fear factor may be even greater than for anthrax.

As of mid-2005, 1,848 cases of hemorrhagic fever in humans caused by Ebola had been reported to the WHO, with 1,287 fatalities (69.9%). Three hundred fifty-four cases of Marburg fever had been reported, with 288 deaths (81.3%). Almost all of these cases arose in Africa. Many have been traced to transmission through unclean clinical syringes, a common problem in rural Africa; the resulting mortality in these cases was 100%. We might hope that mortality would be somewhat lower in industrialized countries, but make no mistake: these viruses are far and away the most lethal pathogens on the CDC's A list.

Both the United States and the Soviet Union produced aerosol versions of HFVs, including Ebola and Marburg viruses, for biological warfare. These were never used, and we have no idea how effective aerosolized HFVs would be. Aerosolized HFVs are relatively stable and cause disease and death in nonhuman primates, and it is presumed they would do the same in humans. Aum Shinrikyo traveled to Zaire to obtain samples of Ebola but was unable to procure enough stock to create a weapon.

Because there have been so few cases, usually occurring in remote areas, exactly what happens in the course of a naturally acquired Ebola or Marburg infection in humans is not entirely clear. The popular depiction of humans being literally melted away from the inside out contains a good deal of dramatic license, but these are undeniably ghastly diseases. Blood vessels as well as blood cells are a frequent target of the viruses and are rapidly destroyed once the virus begins to spread in the body, causing massive internal bleeding. But organs such as the liver and kidneys are also severely damaged. Symptoms of hemorrhagic fever include the usual early signs of any microbial infection, but these are quickly followed by widespread body rashes and blood spots in the skin, blood seepage from various orifices, convulsions, delirium, and a rapid descent into shock and coma. There are no antiviral drugs effective against Marburg and Ebola. For the few people who survive, there is a long period of impairment of numerous body functions.

We know very little about the human immune response to Ebola and Marburg viruses. A 1996 study in Gabon showed that those infected with Ebola who subsequently died of circulatory collapse failed to develop a strong antibody response and had no CD8 killer cell response at all. Among close family members who did not die, about half had produced Ebola antibodies, indicating they had been exposed to the virus, and most also had activated CD8 T cells and a strong inflammatory response.

Why some people were able to mount a protective immune response to Ebola, while others weren't, is not presently known. In mice, we know that both antibodies and CD8 killer cells are induced by exposure to Ebola. Passive transfer of the antibodies to naïve mice did not provide protection against subsequent exposure to Ebola, but transfer of immune CD8 T cells did. Passive transfer of HFV antibodies in nonhuman primates has not generally provided much protection, and there is little hope that this would be an effective treatment in humans.

Attempts to produce an effective vaccine against Ebola and Marburg had been generally unsuccessful until 2005, when a research group centered in Canada developed a single, DNA-based vaccine that is very potent against *both* Ebola and Marburg. Just one injection protected monkeys from infection by either virus. The Ebola/Marburg gene was engineered to be delivered preferentially to macrophages and dendritic cells, to optimize rapid antigen presentation of viral peptides to T cells. It is possible that with further work this vaccine could also be made effective for other A-list HFVs. More work needs to be done before human trials can begin, but this vaccine looks extremely promising.

CAN WE DO IT?

So what do we know about the ability of this wall we hide behind— our immune system—when it comes to bioterrorism agents? Can it help us? Well, the first thing to remember is that, with the agents on the CDC's A list (not to mention lists B and C), if our immune systems could stop these agents dead in their tracks, they wouldn't be on the CDC lists. The real question is, is there anything we can do to help our immune systems do a better job?

One key to helping the immune system is to have the most thorough knowledge possible about how our immune system interacts with these pathogens once they have invaded our bodies. As we have seen, the diseases caused by A-list pathogens are so rare in the United States that we have had little opportunity to study how the immune system responds to them. And until recently we have expended little effort in developing effective clinical measures to guard against them. Most first-line health care responders have no experience with either these pathogens or their diseases, which can cost precious time in identifying the problem in a real attack.

This is quite different from the pathogen that causes AIDS—the HIV virus. We probably know more about every aspect of HIV and its interaction with human beings and their immune systems than any other pathogen on earth. If we are really concerned about bioterrorism with the agents described in this chapter, we need to know much more than we do at present about how they work, and most of all how they are handled by our immune systems. Research programs to answer these questions are currently under way.

The question is often asked, why wouldn't bioterrorists use HIV as a weapon? Why isn't it on the A list? Unquestionably, the release of HIV over a large metropolitan area could generate a maximum fear effect. And as we know all too well, all but a tiny handful of us are defenseless against HIV, with no vaccine on the immediate horizon. So the fear factor probably extends to would-be terrorists themselves. They may be extremely reluctant even to get into the same room with HIV. Another factor is that the incubation period with HIV, before frank AIDS sets in, is 6 to 10 years. Suspiciously large numbers of new cases would likely not be apparent for several years at a minimum. The immediate public relations sensation so craved by terrorists would be lost.

Still, the overall psychological impact on affected populations could be enormous. In the end, the main thing preventing use of HIV is that it is an exceptionally fragile virus. Exposure to anything other than a warm, wet human body disables it within a matter of hours. Aerosolization would almost certainly cripple it. Laboratories working with HIV must take enormous care to keep their strains viable. It is, in fact, a poor candidate for even the CDC's C list.

There are three general strategies for helping our immune systems deal with the kind of pathogens we do find on the A list. The first is to produce enough of what we might call "traditional" vaccines that are effective enough to be of help warding off a bioterrorist attack. Preferably, we would like to have vaccines that could be of help after someone has already been exposed to a particular pathogen. But traditional vaccines are designed to work prophylactically—*before* contact with the pathogen. They are designed to generate protective adaptive immunity in order to boost responses to *subsequent* exposures to the pathogen. Normally, it doesn't matter that several weeks may be required to build up that memory.

Since we don't know which populations of people might be the target of an attack, in order for this approach to be effective we would have to immunize the entire nation—against six pathogens! We do something like that now, with our children, for the most common (and potentially crippling or lethal) childhood infections. And it works. But we do not have vaccines at present for any of the A-list pathogens that are suitable for mass prophylactic immunization programs, for either children *or* adults.

And it is not obvious we would want to undertake such a program even if we had such vaccines. There use would likely be limited to selective immunization of "first responders"—health care personnel, police, fire, certain military units. In the case of anthrax, the causative spores of which could linger in the environment for a long time after a terrorist attack, such vaccines could be useful to immunize individuals present but not infected in an initial attack. A great deal of research is currently directed at making faster-acting vaccines for all the A-list pathogens, and almost certainly we will get some vaccines that induce adaptive immune responses more quickly. But the chances are slim they will be able to act fast enough to be of much use in treating already infected individuals once a bioterrorist attack has been unleashed.

A second approach, still in the largely theoretical stage, takes advantage of the knowledge we have gained over the past decade or so about the workings of the innate immune system. For microbial pathogens (less so for their protein toxins), we now know that the innate immune system plays a direct, cognitive role in the early stages of all infections. Dendritic cells, macrophages, neutrophils, and even B cells have receptors for the pathogen-associated molecular patterns (PAMPs) present on all microbes (chapter 5). The innate immune system, remember, is our first line of defense in any microbial infection, keeping the infection at bay long enough for the adaptive T- and B-cell response to get up and running.

So a great deal of effort is also being expended to find ways to stimulate and strengthen the innate immune response that all of us will be mounting within minutes of any pathogen invasion and lasting for as long as the pathogen remains a threat. The innate response is crucial for triggering inflammation and for processing and presenting forms of microbial antigens that will bring T and B cells into play. Instead of focusing on the antigen-specific, adaptive aspects of vaccination, more attention is being placed on "vaccines" that provide as strong a stimulus as possible to pumping up the critical innate elements of the immune system at the beginning of the response. And because they are not directed at any particular agent, we would need only one such vaccine.

An incoming pathogen will of course trigger these responses on its own, but if substances can be quickly introduced into the body to accelerate that portion of the immune response, it is reasoned, we may be able to stave off the deadliest aspects of infection long enough for the more potent adaptive immune response to get off the ground. The studies of survivors of Ebola and Marburg outbreaks referred to in the last chapter provide strong motivation for this general approach.

The third approach to helping the immune system is to build better antibodies to use for passive immunization. Ready-made antibodies provide a powerful weapon against any bacterial and many viral infections and are also useful for neutralizing microbial toxins. If injected during the first 24 hours or so after someone is infected with a pathogen or toxin, infection could in many instances be enormously reduced, buying precious time for the innate system to complete its job and for the adaptive system to begin functioning.

The problem, as we have seen previously, is that most of these antibodies are made in animals, usually horses, and the antibodies themselves trigger an immune response in the person into whom they are injected. A single injection of horse antibodies into a person doesn't cause a problem, because by the time that person makes antibodies against the incoming horse antibodies, the horse antibodies are gone. Those not taking part in neutralizing microbes are cleared from the blood, like any other protein. But a subsequent administration of horse antitoxin antibodies into that same person would quickly encounter large quantities of that person's antihorse antibodies and be neutralized. And it is entirely possible in a terrorist attack with large amounts of a deadly pathogen that one injection of antibody might not be enough.

Humans recovering from natural infections or planned immunizations with crippled pathogens are also a source for antibodies
that could be used for passive immunization. Although a slight
immune response would be triggered in the person receiving these
antibodies, the response would be relatively mild compared to the
reaction against horse antibodies and could be managed. But the
number of persons from whom such antibodies could be harvested
is vanishingly small compared to the huge numbers of people who
might need treatment in the immediate aftermath of a bioterrorist
attack with a particular pathogen.

Great effort is now being directed at producing "humanized"
antimicrobial antibodies for use in passive immunization. This
approach depends on the technique of monoclonal antibodies
described in chapter 2, with a little genetic razzle-dazzle thrown
in. Antibodies, say, to *B. anthracis* would first be produced in mice.
Mouse B cells producing this antibody would then be isolated and
converted to monoclonal B cells, which can be expanded enormously and used to produce theoretically unlimited amounts of
monoclonal antibody specific for *B. anthracis*.

But these are still mouse antibodies. They will trigger the same
kind of vigorous immune response in humans that horse antibodies do. This is where the razzle-dazzle comes in. It is possible to
genetically engineer mouse B cells so that they produce monoclonal
antibodies with most of their mouse portions replaced with a
human counterpart (Figure 14.1). These humanized mouse antibodies will provoke a greatly reduced immune response in human
recipients, one that will not wipe out a subsequent administration
of humanized antibody. Initial trials of this concept in animal
models have been highly encouraging.

So there is hope! Our immune systems clearly will need some
help in building an effective immune response to pathogens used
in a terrorist attack. Once our immune responses have a chance to
get off the ground and make it to the adaptive response stage, they
will be more than able to defend us against not only the first attack with a given pathogen, but any subsequent exposures as well.

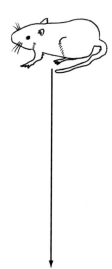

A mouse is genetically engineered so that its antibody heavy and light chain constant region gene fragments have been exchanged with human H and L C-region gene fragments. (See also Figure 2.2.)

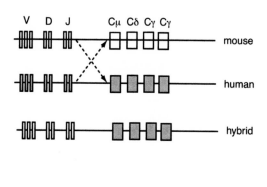

This mouse will produce antibodies that have mouse V regions, but human C regions. Most of what a human sees as foreign in mouse antibodies is contained in their C regions, so these antibodies will look mostly human to a human, and will provoke a much milder immune response.

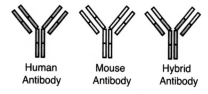

FIGURE 14.3
Creation of "humanized" mouse monoclonal antibodies.

It may be that none of the three approaches just described, by themselves, can provide the help we need, but by combining them in the way we combine different approaches to treating cancer, there is a very good chance that we can gain the most important thing we need to respond effectively to a bioterrorist attack—time!

Glossary

Adaptive immunity Immunity directed toward specific antigenic epitopes (q.v.) rather than generalized toxicity toward microbes

Aerosol Form of a substance that can be disseminated on air currents

Anaphylactic shock Physiologic shock arising secondary to an immunological overreaction, usually involving IgE and mast cells

Agammaglobulinemia Condition in which an individual is unable to make antibodies

Agglutinate To clump

Allergen Foreign antigen capable of inducing an allergic reaction

Allograft A graft exchanged between two genetically nonidentical members of the same species

Angiogenesis The generation of new blood vessels

Antibody A protein found in blood, produced in response to invasion of the body by a microbe or other foreign biological entity, capable of recognizing that entity and promoting its elimination

Antigen A foreign biological entity capable of inducing production of an antibody and reacting in a chemically specific fashion with the inducing antibody

Antigen processing Breakdown of protein antigens by dendritic cells into smaller peptides, which are then associated with major histocompatibility complex (MHC) molecules for presentation to T cells

Antigen receptor   Molecule on the surface of a T or B cell that interacts with foreign antigen, leading to activation of the cells and elimination of the antigen

Antigenic epitope   The restricted portion of a large, complex antigen with which an antibody specifically interacts

Antiserum   Serum from an immunized animal or person, containing antibodies against the substance used for immunization

Apoptosis   A form of cell suicide, also called "programmed cell death"

Attenuated pathogen   A pathogen disabled with respect to its ability to induce disease, but still able to induce immunity against itself

Autograft   A graft transplanted from one part of an individual to another part of the same individual

Autoimmunity   Immune responses directed toward self molecules and tissues

B cell   A lymphocyte produced in the bone marrow and residing in lymph nodes and spleen, responsible for production of antibodies

Bacillus   Rod-shaped bacterium

Basophil   White blood cell having many of the properties of mast cells (q.v.)

Bone marrow   A soft, jelly-like substance found at the center of most bones in the body; contains hematopoietic stem cells from which all mature blood cells derive

Botox   Botulinum toxin used for medicinal purposes

Buboes   Obsolete term for enlarged lymph nodes

Chemokine   A cytokine (q.v.) released by one cell that functions to attract other cells to the same location

Cirrhosis   Degenerative disease of the liver

Complement   A series of proteins that mediate numerous immune functions such as inflammation, cell lysis, and tagging for phagocytosis

Cytokine   Chemical message used by cells to communicate with each other

Dendritic cell A bone marrow–derived white cell that senses the presence of microbes, releases inflammatory cytokines, and processes antigen for presentation to T cells

DNA vaccine A vaccine in which microbial DNA is used, rather than the microbe itself. The host produces microbial proteins from the DNA, which then induce an immune response. DNA vaccines are very stable.

DTH Delayed-type hypersensitivity

EBV Epstein-Barr virus, involved in mononucleosis and lymphoma. It is a common opportunistic pathogen

Edema Local swelling resulting from failure to drain lymph fluid from tissue

Embryonic stem cells Cells taken from the early stages of fetal development, which retain the ability under special conditions to reproduce an entire adult being

Epitope See *antigenic epitope*

Erythema Skin rash caused by capillary enlargement

Erythematosus A type of skin rash

Erythrocyte A red blood cell

Extravasation Escape of white blood cells from a blood vessel into the surrounding tissue

Fc receptor Structure found on various white blood cells that allows them to bind to the tail portion of an antibody molecule

Frank AIDS Clinically defined AIDS, as opposed to HIV infection

Freemartin Fraternal twins sharing the same placenta

Gamma globulin The subset of blood proteins to which antibodies belong

Gene therapy Use of DNA containing functional genes to correct genetic defects

Genome The total set of DNA sequences defining an individual or species

Glomerulonephritis Inflammation of the filtering tubules of the kidneys; often results from deposition of antigen–antibody complexes formed in the course of autoimmune disease

Granuloma   Nodules of white blood cells that sometimes form in response to a local microbial infection

GVH reaction   Graft versus host reaction, in which white cells grafted into an immunoincompetent host attack host cells

Hematopoiesis   The generation of blood cells (both red and white) from a hematopoietic stem cell in bone marrow

Hemorrhagic fever   Inflammatory state, usually virally induced, accompanied by excessive bleeding

Hepatitis   Inflammation of the liver

Histocompatibility   Degree of genetic relatedness of cells from different sources

Host   Animal harboring parasitic organisms

Hybridoma   A hybrid cell made by fusing a normal cell with a tumor cell. Under the right conditions, the hybrid will retain the unlimited growth potential of the tumor cell together with the specialized function of the normal cell.

Hypersensitivity   An excessive immune response that can cause tissue damage

Immune complex   Aggregated antibodies and antigen

Immunodeficiency   Condition in which one or more parts of the immune system are missing, diseased, or defective

Immunogenic   Capable of inducing an immune response

Immunoglobulin   Technical term for antibodies, reflecting their origin in the gamma globulin fraction of blood proteins

Immunopathology   Disease caused by the immune system

Inbred animals   Animals produced by repeated brother–sister matings, which after 20 or so generations become essentially genetically identical

Infectious disease   Disease caused by infectious, pathogenic microbes

Inflammation   A response generated primarily by the innate immune system (q.v.) involving changes in blood flow, release of cytokines, and altered white cell trafficking

Innate immunity   Immunity directed against a generalized threat, such as microbial invasion, without recognizing any microbe-specific antigenic structures

Interferons   Chemicals produced by cells to interfere with viral infection

Isograft   A graft exchanged between two genetically identical individuals

Leukocyte   A white blood cell (q.v.)

Lipophilic   Fat-like in terms of chemical nature

Lymph   Fluid that circulates from tissue to blood and back again, carrying food, oxygen, and waste products as well as cells of the immune system (See lymphatic vessels)

Lymph node   Sac-like structures placed along lymphatic circulatory vessels; involved in trapping antigen, and home to numerous cells of the immune system

Lymphatic vessels   Pick up fluid from tissue spaces, fusing into ever-larger vessels that return the fluid to the blood at the great veins of the neck

Lymphocyte   A subset of white blood cells concerned with mediating adaptive immunity

Lymphoma   Cancer involving white blood cells

Macrophage   A large cell specialized for phagocytosis (q.v.)

Mast cells   Histamine-producing white blood cells often involved in allergic reactions

Memory   With respect to adaptive immunity, refers to the changes wrought in T and B cells as a result of encounter with antigen, enabling them to respond more efficiently upon subsequent encounter with antigen

MHC   Major histocompatibility complex, the collection of genes controlling a wide range of immune reactions

Microbe   A single-cell living organism

Monoclonal antibodies   Antibodies derived from the progeny of a single ancestral B cell, and thus all identical

Neonatal   Newborn

Neutrophil   White blood cell that can ingest and destroy microbes

NK cell   Natural killer cell, a white blood cell of the innate immune system involved in controlling cancer and viral infections

Oncogene   A gene involved in initiating cell division, which, in mutant form, may trigger unscheduled cell division and contribute to cancer

Opportunistic pathogen   A pathogen that lives within a host without causing disease, except when host immune defenses are impaired

Pan-autoimmunity   A condition in which a patient experiences autoimmune disease in a number of different tissue sites

Passive immunization   Transfer of preformed antibodies or T cells from a person with immunity against a given pathogen to a person lacking such immunity

Pathogenic   Causing disease

Phagocytosis   Engulfment of a cell or other biological material by a specialized cell called a phagocyte. The ingested material is degraded, and portions of it may be displayed on the surface of the phagocyte.

Plasma   Serum that still retains blood-clotting factors

Polymorphism   Literally, "multiformed." With respect to antibodies, it refers to the multiple forms of antibody generated within an individual. With respect to major histocompatibility complex (MHC) genes and proteins, it refers to the multiple forms found within a species.

Protozoa   Single-cell life forms, one step up in complexity from bacteria, the simplest single-cell life form

Reservoir   In epidemiology, an animal that can harbor a human pathogen without itself contracting the disease caused by that pathogen in humans

Retrovirus   An RNA virus that copies its RNA into DNA, which it then inserts into the host DNA genome

Rhinitis   Inflammation of the nose

SCID   Severe combined immunodeficiency disease, which arises through an inherited genetic defect in T cells

Sepsis Condition in which microbes (usually bacteria) replicate within the body with no control by the immune system

Septicemia Sepsis restricted to the blood system

Seroconversion The first appearance of HIV antibodies in the blood after infection with HIV

Serum The fluid portion of blood remaining after red blood cells have been clotted and removed

Spore Bacterial state of suspended animation induced by starvation that reverses when food and water are available

Stem cells Progenitor cells that divide to produce one stem cell daughter and one daughter destined to become a particular adult cell

T cell A thymus-derived lymphocyte that produces cytokines and promotes or regulates other immune-cell functions

Thymectomy Surgical removal of the thymus

Thymus Organ situated just above the heart, concerned with maturation of T lymphocytes

Tissue typing Determining the degree of major histocompatibility complex (MHC) relatedness of two individuals

TNF Tumor necrosis factor, a cytokine produced by certain white cells

Tolerance In the immune system, making sure antibodies or T cells recognizing self components are either eliminated or brought under tight control. Failure of tolerance can result in autoimmunity.

Toxin In microbiology, a chemical released by a parasitic microbe that is harmful to its host

Toxoid Chemically altered toxin molecule retaining its immunogenicity

Transfection With respect to DNA, the introduction of DNA from an external source into a cell

Tuberculin Protein extract of *Mycobacterium tuberculosis*

Vaccination Immunization with all or part of a pathogen, intended to build immunity before natural exposure to the pathogen

Virgin   With respect to T and B lymphocytes, refers to cells that have not yet encountered antigen

Virion   An individual virus particle

Virulence   Disease-evoking power of a pathogenic microbe

White blood cells   Unpigmented cells of various hematopoietic lineages found in the blood, after red blood cells have been removed. All play a role in various immune phenomena.

Xenograft   A graft exchanged between members of different species

Index

CPSIA information can be obtained at www.ICGtesting.com
Printed in the USA
BVOW082357080212

282478BV00001B/2/P